Measuring Lead Exposure in Infants, Children, and Other Sensitive Populations

COMMITTEE ON MEASURING LEAD
IN CRITICAL POPULATIONS

BOARD ON ENVIRONMENTAL STUDIES
AND TOXICOLOGY

COMMISSION ON LIFE SCIENCES

NATIONAL ACADEMY PRESS
WASHINGTON, D.C. 1993

Tennessee Tech Library
Cookeville, TN

NATIONAL ACADEMY PRESS   2101 Constitution Ave., N.W. Washington, D.C. 20418

NOTICE: The project that is the subject of this report was approved by the Governing Board of the National Research Council, whose members are drawn from the councils of the National Academy of Sciences, the National Academy of Engineering, and the Institute of Medicine. The members of the committee responsible for the report were chosen for their special competencies and with regard for appropriate balance.

This report has been reviewed by a group other than the authors according to procedures approved by a Report Review Committee consisting of members of the National Academy of Sciences, the National Academy of Engineering, and the Institute of Medicine.

The National Academy of Sciences is a private, non-profit, self-perpetuating society of distinguished scholars engaged in scientific and engineering research, dedicated to the furtherance of science and technology and to their use for the general welfare. Upon the authority of the charter granted to it by the Congress in 1863, the Academy has a mandate that requires it to advise the federal government on scientific and technical matters. Dr. Bruce M. Alberts is president of the National Academy of Sciences.

The National Academy of Engineering was established in 1964, under the charter of the National Academy of Sciences, as a parallel organization of outstanding engineers. It is autonomous in its administration and in the selection of its members, sharing with the National Academy of Sciences the responsibility for advising the federal government. The National Academy of Engineering also sponsors engineering programs aimed at meeting national needs, encourages education and research, and recognizes the superior achievements of engineers. Dr. Robert M. White is president of the National Academy of Engineering.

The Institute of Medicine was established in 1970 by the National Academy of Sciences to secure the services of eminent members of appropriate professions in the examination of policy matters pertaining to the health of the public. The Institute acts under the responsibility given to the National Academy of Sciences by its congressional charter to be an adviser to the federal government and, upon its own initiative, to identify issues of medical care, research, and education. Dr. Kenneth I. Shine is president of the Institute of Medicine.

The National Research Council was organized by the National Academy of Sciences in 1916 to associate the broad community of science and technology with the Academy's purposes of furthering knowledge and advising the federal government. Functioning in accordance with general policies determined by the Academy, the Council has become the principal operating agency of both the National Academy of Sciences and the National Academy of Engineering in providing services to the government, the public, and the scientific and engineering communities. The Council is administered jointly by both Academies and the Institute of Medicine. Dr. Bruce M. Alberts and Dr. Robert M. White are chairman and vice chairman, respectively, of the National Research Council.

The project was supported by the Comprehensive Environmental Response, Compensation, and Liability Act Trust Fund through cooperative agreement with the Agency for Toxic Substances and Disease Registry, U.S. Public Health Service, Department of Health and Human Services.

Library of Congress Catalog Card No. 93-84436
International Standard Book No. 0-309-04927-X

B-152

Copyright 1993 by the National Academy of Sciences. All rights reserved.

Printed in the United States of America

Committee on Measuring Lead in Critical Populations

BRUCE A. FOWLER *(Chairman)*, University of Maryland School of Medicine, Baltimore, Md.
DAVID C. BELLINGER, Children's Hospital, Boston, Mass.
ROBERT L. BORNSCHEIN, University of Cincinnati Medical Center, Cincinnati, Ohio
J. JULIAN CHISOLM, The Kennedy Institute, Baltimore, Md.
HENRY FALK, Centers for Disease Control, Atlanta, Ga.
A. RUSSELL FLEGAL, University of California, Santa Cruz, Calif.
KATHRYN R. MAHAFFEY, National Institute of Environmental Health Sciences and University of Cincinnati Medical School, Cincinnati, Ohio
PAUL MUSHAK, University of North Carolina, Durham, N. Car.
JOHN F. ROSEN, Montefiore Medical Center, Bronx, N.Y.
JOEL SCHWARTZ, U.S. Environmental Protection Agency, Washington, D.C., and Harvard School of Public Health, Cambridge, Mass.
RODNEY K. SKOGERBOE (retired), Colorado State University, Loveland, Colo.

Technical Advisers

GEORGE PROVENZANO, University of Maryland, Baltimore, Md.
JOEL POUNDS, Institute of Chemical Toxicology, Wayne State University, Detroit, Mich.

Staff

RICHARD D. THOMAS, Program Director
CAROLYN E. FULCO, Staff Officer (until June 1990)
MARY B. PAXTON, Staff Officer (until April 1991)
NORMAN G. GROSSBLATT, Editor
RUTH E. CROSSGROVE, Information Specialist and Copy Editor
SHELLEY A. NURSE, Senior Project Assistant
RUTH P. DANOFF, Project Assistant

Sponsor: Agency for Toxic Substances and Disease Registry

Board on Environmental Studies and Toxicology

PAUL G. RISSER *(Chair)*, University of Miami, Oxford, Ohio
FREDERICK R. ANDERSON, Cadwalader, Wickersham & Taft, Washington, D.C.
MICHAEL J. BEAN, Environmental Defense Fund, Washington, D.C.
EULA BINGHAM, University of Cincinnati, Cincinnati, Ohio
EDWIN H. CLARK, Department of Natural Resources and Environmental Control, State of Delaware, Dover, Del.
ALLAN H. CONNEY, Rutgers University, N.J.
JOHN L. EMMERSON, Eli Lilly & Company, Greenfield, Ind.
ROBERT C. FORNEY, Unionville, Pa.
ROBERT A. FROSCH, Harvard University, Cambridge, Mass.
ALFRED G. KNUDSON, Fox Chase Cancer Center, Philadelphia, Pa.
KAI LEE, Williams College, Williamstown, Mass.
JANE LUBCHENCO, Oregon State University, Corvallis, Ore.
HAROLD A. MOONEY, Stanford University, Stanford, Calif.
GORDON ORIANS, University of Washington, Seattle, Wash.
FRANK L. PARKER, Vanderbilt University, Nashville, Tenn., and Clemson University, Anderson, S. Car.
GEOFFREY PLACE, Hilton Head, S. Car.
DAVID P. RALL, Washington, D.C.
LESLIE A. REAL, Indiana University, Bloomington, Ind.
KRISTIN SHRADER-FRECHETTE, University of South Florida, Tampa, Fla.
BAILUS WALKER, JR., University of Oklahoma, Oklahoma City, Okla.
GERHARDT ZBINDEN, Eidgenossische Technische Hochschule Zurich, Schwerzenbach, Switzerland

Staff

JAMES J. REISA, Director

DAVID J. POLICANSKY, Associate Director and Program Director for Natural Resources and Applied Ecology

RICHARD D. THOMAS, Associate Director and Program Director for Human Toxicology and Risk Assessment

LEE R. PAULSON, Program Director for Information Systems and Statistics

RAYMOND A. WASSEL, Program Director for Environmental Sciences and Engineering

Commission on Life Sciences

THOMAS D. POLLARD *(Chair)*, Johns Hopkins Medical School, Baltimore, Md.
BRUCE M. ALBERTS, University of California, Berkeley, Calif.
BRUCE N. AMES, University of California, Berkeley, Calif.
J. MICHAEL BISHOP, Hooper Research Foundation, University of California Medical Center, San Francisco, Calif.
DAVID BOTSTEIN, Stanford University School of Medicine, Stanford, Calif.
MICHAEL T. CLEGG, University of California, Riverside, Calif.
GLENN A. CROSBY, Washington State University, Pullman, Wash.
LEROY E. HOOD, University of Washington, Seattle, Wash.
MARIAN E. KOSHLAND, University of California, Berkeley, Calif.
RICHARD E. LENSKI, University of Oxford, Oxford, United Kingdom
STEVEN P. PAKES, Southwestern Medical Center, Dallas, Tex.
EMIL A. PFITZER, Hoffmann-La Roche Inc., Nutley, N.J.
MALCOLM C. PIKE, University of Southern California School of Medicine, Los Angeles, Calif.
PAUL G. RISSER, Miami University, Oxford, Ohio
JOHNATHAN M. SAMET, University of New Mexico School of Medicine, Albuquerque, N.Mex.
HAROLD M. SCHMECK, JR., Armonk, N.Y.
CARLA J. SHATZ, University of California, Berkeley, Calif.
SUSAN S. TAYLOR, University of California at San Diego, La Jolla, Calif.
P. ROY VAGELOS, Merck and Company, Inc., Whitehouse Station, N.J.
TORSTEN N. WIESEL, Rockefeller University, New York, N.Y.

PAUL GILMAN, Executive Director

Other Recent Reports of the Board on Environmental Studies and Toxicology

Pesticides in the Diets of Infants and Children (1993)
Issues in Risk Assessment (1993)
Setting Priorities for Land Conservation (1993)
Protecting Visibility in National Parks and Wilderness Areas (1993)
Biologic Markers in Immunotoxicology (1992)
Dolphins and the Tuna Industry (1992)
Environmental Neurotoxicology (1992)
Hazardous Materials on the Public Lands (1992)
Science and the National Parks (1992)
Animals as Sentinels of Environmental Health Hazards (1991)
Assessment of the U.S. Outer Continental Shelf Environmental Studies Program, Volumes I-IV (1991-1993)
Human Exposure Assessment for Airborne Pollutants (1991)
Monitoring Human Tissues for Toxic Substances (1991)
Rethinking the Ozone Problem in Urban and Regional Air Pollution (1991)
Decline of the Sea Turtles (1990)
Tracking Toxic Substances at Industrial Facilities (1990)
Biologic Markers in Pulmonary Toxicology (1989)
Biologic Markers in Reproductive Toxicology (1989)

*Copies of these reports may be ordered from
the National Academy Press
(800) 624-6242
(202) 334-3313*

Preface

Lead is a ubiquitous toxicant. It is especially toxic to young children and the fetus. Evidence gathered recently has shown that lead concentrations of less than half the previous Centers for Disease Control (CDC) guideline (30 μg/dL) can impair cognitive and physical development in children and increase blood pressure in adults. In response to that evidence, the U.S. Environmental Protection Agency has proposed setting 10 μg/dL as a maximal blood lead concentration. On the basis of the same evidence of toxic effects at 10 μg/dL, CDC has recently reduced its 1985 intervention or action concentration from 25 μg/dL to 10 μg/dL and proposed a concentration for clinical management of 20 μg/dL.

Persons exposed to lead and with blood lead concentrations above 10 μg/dL are likely to number in the millions. It was estimated that in 1984 about 6 million children and 400,000 fetuses in the United States were exposed to lead at concentrations that placed them at risk of adverse health effects (i.e., blood lead concentrations of at least 10 μg/dL). Because of the potential for toxic exposures to lead of a large segment of the population, especially sensitive populations (infants, children, and pregnant women), there is a need to develop and refine methods for measuring lead in blood at the revised lower concentrations (10 μg/dL).

In addition, new reliable and reproducible techniques for measuring lead in other tissues, such as bone, will also need to be developed. Methods for detecting and measuring biologic markers of low-dose exposure to lead are also needed, because the erythrocyte protoporphyrin test lacks sensitivity at blood lead concentrations below 25 μg/dL. The U.S. Agency for Toxic Substances and Disease Registry (ATSDR) requested that the National Research Council (NRC) provide information on measuring environmental exposure of sensitive populations to lead. In response, the Board on Environmental Studies and Toxicology in the NRC Commission on Life Sciences formed the Committee on Measuring Lead Exposure in Critical Populations, which produced this report.

Committee members have expertise in toxicology, epidemiology, medicine, and chemistry. From the beginning, the committee decided to take a broad view of its charge and to produce a report that would not only consider a variety of technical methods for measuring lead and biologic markers of lead exposure in human populations at special risk for lead toxicity, but would consider related issues, such as sources of exposures and toxicity in sensitive populations.

We hope that this document meets the goals of ATSDR, which took the initiative in sponsoring this study, and the needs of the wide array of readers and regulators concerned with the impact of lead toxicity in human populations at special risk. It is clear that public-health problems associated with the misuse of lead have plagued society for several thousand years. Modern humans are estimated to have total body burdens of lead approximately 300-500 times those of our prehistoric ancestors, because lead is extensively mobilized from the earth's crust by our activities. This committee believes that the state of scientific knowledge and technical tools to deal with the lead problem are sufficiently developed to begin the process of changing these public health risks. We hope that this report will be a useful tool to those charged with shaping effective approaches for dealing with lead toxicity and thus improving the public health.

The committee gratefully acknowledges the interest and support of Barry Johnson of ATSDR. We also thank George Provenzano, University of Maryland, Baltimore, and Joel Pounds, Institute of Chemical Toxicology, Wayne State University, who provided information for the committee. Finally, the committee was concerned about the extent to which societal resources are necessary to implement various environmental lead control options that are associated with the analytical methods described in this report. The committee requested that one of its members, Joel Schwartz, prepare a detailed benefit analysis of lead exposure prevention. His analysis will appear in the *Journal of Environmental Research*. The committee is grateful for his independent analysis of this important policy issue.

This report could not have been produced without the untiring efforts of Shelley Nurse, senior project assistant. Norman Grossblatt edited the report. Finally, the committee gratefully acknowledges the persistence, patience, and expertise of Carolyn E. Fulco, project director for the study until June 1990; Mary B. Paxton, project director until April 1991; and Richard D. Thomas, project director from May 1991. Dr. Thomas, an expert in toxicology and public health, provided us the guidance, perspective, and judgmental interventions necessary to bring this report to its final form.

      Bruce A. Fowler
      Chairman
      Committee on Measuring Lead
        in Critical Populations

Contents

EXECUTIVE SUMMARY ... 1

1 INTRODUCTION ... 13
  Perspective on Issues, 14
  Historical Background, 22
  Scope and Organization of the Committee Report, 29

2 ADVERSE HEALTH EFFECTS OF EXPOSURE TO LEAD ... 31
  Clinical Intoxication in Children, 32
  Intoxication in Adults, 34
  Reproductive and Developmental Effects, 35
  Cardiovascular Effects, 72
  Mechanisms of Toxicity, 77
  Summary, 93

3 LEAD EXPOSURE OF SENSITIVE POPULATIONS ... 99
  Historical Overview of Anthropogenic
    Lead Contamination, 100
  Source-Specific Lead Exposure of
    Sensitive Populations, 109
  Summary, 139

4 BIOLOGIC MARKERS OF LEAD TOXICITY ... 143
  Biologic Markers of Exposure, 143
  Biologic Markers of Effect, 167

(CHAPTER 4, CONT.)
    Biologic Markers of Susceptibility, 183
    Summary, 187

5   METHODS FOR ASSESSING EXPOSURE TO LEAD   191
    Introduction, 191
    Sampling and Sample Handling, 196
    Measurement of Lead in Specific Tissues, 197
    Mass Spectrometry, 205
    Atomic-Absorption Spectrometry, 218
    Anodic-Stripping Voltammetry and Other
      Electrochemical Methods, 223
    Nuclear Magnetic Resonance Spectroscopy, 224
    The Calcium-Disodium EDTA Provocation Test, 226
    X-Ray Fluorescence Measurement, 227
    Quality Assurance and Quality Control, 244
    Summary, 247

6   SUMMARY AND RECOMMENDATIONS   253
    Sources of Lead Exposure, 253
    Adverse Health Effects of Lead, 254
    Markers of Lead Exposure and Effect, 256
    Techniques to Measure Lead Exposure
      and Early Toxic Effects, 257

REFERENCES   263

Measuring Lead Exposure in Infants, Children, and Other Sensitive Populations

Executive Summary

Adverse health effects from exposure to lead are now recognized to be among industrialized society's most important environmental health problems. In the United States, more than 6 million preschool metropolitan children and 400,000 fetuses were believed to have lead concentrations above 10 micrograms per deciliter (μg/dL) of whole blood. That concentration has been designated by the U.S. Public Health Service as the maximum permissible concentration from the standpoint of protecting the health of children and other sensitive populations, and 20 μg/dL is the concentration at which medical intervention should be considered. A blood lead concentration of 10 μg/dL is low by comparison with the concentrations that have been associated with observable toxic reactions and that used to be widely permitted in the 1970s (e.g., 50-80 μg/dL). But it is hundreds of times *higher* than estimated blood lead concentrations in preindustrial peoples. For example, studies of bone samples of North American Indians and other preindustrial populations indicate that body burdens of lead in the general population today are 300-500 times greater than preindustrial background concentrations.

Science and society have been remarkably slow to recognize and respond to the full range of harm associated with lead exposure, but that is changing. Understanding of this public-health problem has involved a complex mixture of scientific knowledge, societal perception of risk, economic concerns based on lead's numerous uses, and a recognition that adverse health effects of lead are associated with lower exposures than previously believed. Current scientific studies in toxicology and epidemiology have shown that relatively low blood concentrations of lead may be associated with toxic effects. The appropriate regulatory agencies have responded or have begun to respond by lowering the allowable (i.e., presumably safe) concentrations of lead in the population.

Until the early 1970s in the United States, the acceptable concentration of lead in blood was 60 μg/dL in children and 80 μg/dL in adults. Early in 1990, after a series of intermediate lowerings of the acceptable concentration by various agencies and organizations, the Science Advisory Board of the U.S. Environmental Protection Agency (EPA) identified a blood lead concentration of 10 μg/dL as the maximum to be considered safe for individual young children, on the basis of available evidence. The U.S. Centers for Disease Control and Prevention (CDC) also recently lowered its lead-exposure guideline to 10 μg/dL and its guideline for medical intervention to 20 μg/dL. It should be noted that CDC, in a 1985 statement, explicitly identified lead toxicity in children at blood lead concentrations well below 25 μg/dL, but a lower concentration was not chosen as a guideline at that time, because of the logistics and feasibility of lead screening. As with the previous reduction in exposure limits, the advent of more sensitive and reliable analytic techniques has played a central role in these changes in permissible exposures.

CHARGE TO THE COMMITTEE AND STRUCTURE OF THE REPORT

This report was prepared by the National Research Council's Committee on Measuring Lead Exposure in Critical Populations, a committee of the Board on Environmental Studies and Toxicology. The study was sponsored by the Agency for Toxic Substances and Disease Registry (ATSDR).

As part of its efforts to prepare this report, the committee summarized new scientific evidence on the low-dose toxicity of lead, and generally concurs with CDC in the selection of 10 μg/dL as the concentration of concern in children. Meeting this new guideline will require substantial improvement in methods for measuring lead in blood and monitoring other biologic markers of lead in tissues. If preventive techniques are to be successful, amounts and sources of lead must be identified. Developing analytic measurement techniques that are accurate and precise at such low concentrations is a difficult scientific challenge.

EXECUTIVE SUMMARY

The committee was charged to examine one segment of the lead issue: the assessment of lead exposure in sensitive populations via various biologic markers of exposure and early effects. Chapters 2 and 3 of this report summarize the toxicity of lead and sources of exposure to lead for sensitive populations, defined in this report as infants, children, and pregnant women. Chapter 4 deals with lead in blood and other physiologic media and describes the monitoring of biologic markers that indicate that exposure to lead has occurred, markers of early toxic effects, and markers of susceptibility. Chapter 5 assesses techniques for quantitative measurement of the biologic markers of exposure and effect; it concludes by describing trends in monitoring lead exposure and the effects on society of reducing exposure. Finally, Chapter 6 presents the committee's conclusions and recommendations to improve the monitoring of lead in sensitive populations.

CONCLUSIONS

Sensitive Populations

As CDC has concluded, blood lead concentrations at or around 10 μg/dL present a public-health risk to sensitive populations on the basis of current evidence. The sensitive populations with respect to these adverse effects are infants, children, and pregnant women (as surrogates for fetuses). There is growing evidence that even very small exposures to lead can produce subtle effects in humans. Therefore there is the possibility that future guidelines may drop below 10 μg/dL as the mechanisms of lead toxicity become better understood.

The adverse effects noted at approximately 10 μg/dL include

- Impairments of CNS and other organ development in fetuses.
- Impairments in cognitive function and initiation of various behavioral disorders in young children.
- Increases in systolic and diastolic blood pressure in adults including pregnant women.
- Impairments in calcium function and homeostasis in sensitive populations found in relevant target organ systems.

Somewhat higher blood lead concentrations are associated with impairment in biosynthesis of heme, a basic substance required for blood formation, oxygen transport, and energy metabolism. Some effects described above—cognitive dysfunction and behavioral disorders—might well be irreversible. One recent study showed persistence into early adulthood of childhood neurobehavioral effects due to lead. The revelation of adverse effects after modest exposures in study populations forces the question of what the aggregate impact on sensitive populations is, with respect to current exposures.

Quantitative Methods for Analysis

Exposure

It is estimated that millions of infants, children, and pregnant women in the United States have blood lead concentrations above 10 μg/dL. And, the toxicity of lead is established across the spectrum of exposure concentrations starting as low as 10 μg/dL. The exposure span between these low-dose effects and the concentrations of lead associated with substantial risk of severe brain damage and death is a factor of only 10-15 for a child and probably less for a fetus. In contrast, safety margins of 10-100 for other toxic substances are commonly used with the lowest-observed-effect level (LOEL) in humans to establish an acceptable exposure of the general population.

Sources and Accumulation

With the reduction of lead in gasoline and foods, the remaining major sources of lead are

- Lead-based paint.
- Dust and soil.
- Drinking water.
- Occupational exposure.

Lead-based paint is the largest source of high-dose lead exposure for children. Dust and soil can be high-dose sources and can also constitute important sources of general population exposure. Drinking water is also a major source for the general population, and sometimes high concentrations of lead are found in drinking water. Lead in food is primarily a general population source. Although lead in food declined markedly during the 1980s, primarily because of the decrease in use of lead-soldered cans manufactured in the United States, imported canned foods continue to be high in lead. Gasoline lead was the major source of general population exposure in the 1970s, but regulatory action has reduced it by over 95%.

Dust and soil lead is a legacy of past production of lead, as well as past uses in paint, gasoline, and other substances. Dust and soil lead continues to be replenished by the deterioration of lead-based paint and other sources. It serves as a compelling environmental reminder that lead is not biodegradable and will accumulate in areas with substantial loadings. In addition, stationary sources of lead, such as smelters, can be regionally important.

RECOMMENDATIONS

Sensitive Populations

The committee had as a principal task the characterization of members of the population who are at increased risk for lead exposure and toxicity and are therefore members of a "sensitive" population. The committee identified a number of sensitive populations in which low-dose lead exposure assessments were necessary. They include infants, children, and pregnant women. Other populations are at risk because of potentially large exposures. The committee concluded, however, that the most sensitive populations are infants, children, and pregnant women; these populations are the focus of this report. (Lead workers have long been recognized to be at high risk because of excessive exposure.)

Populations are defined as sensitive according to intrinsic and extrinsic factors or mixtures of the two. Age, sex, and genetic susceptibilities typify intrinsic factors; relationships of subjects to external exposure sources define extrinsic factors. Mixtures of the two can exist, as in female workers who are exposed to lead in the workplace.

Quantitative Methods for Analysis

The principal markers of exposure are lead concentrations in various physiologic media, of which whole blood is the most commonly used for exposure assessment in sensitive populations. In very young children, lead in whole blood is an indicator mainly of recent exposure although there can be variable (but not dominant) input to total blood lead from past accumulation in the body. In adults and particularly lead workers, the past accumulation is a more prominent contributor to total blood lead. However, the historical input is determined by the slow kinetic component in blood-lead decay rates and thus, it is rarely the dominant contributor to total blood lead.

Requirements for a longer-term measure of continuing lead exposure in sensitive populations necessitates use of in vivo measurements of lead in bone. Lead in shed teeth reflects lead accumulation over the period from tooth eruption to shedding in children and is useful for quantifying accumulation, but is inadequate as a basis for regulatory action, because it reflects exposure over a long period.

Traditionally, impairments of steps in the biosynthesis of heme have been exploited as effect markers in sensitive populations. The accumulation of erythrocyte protoporphyrin (EP or ZPP) in whole blood was once the primary screening test to identify children with increased lead burdens. As blood lead concentrations of concern have continued to decline, however, this measure does not retain the necessary sensitivity or the specificity. One measure judged not to have meaning for systemic toxicity is inhibition in activity of porphobilinogen synthetase (ALA-dehydratase), an enzyme in the erythrocyte. Experimental animal studies have identified lead-binding proteins and stoichiometrically interactive processes of lead with calcium systems in various tissues. Their immediate relevance or feasibility for routine exposure screening in sensitive populations remains undefined.

The committee recommends

EXECUTIVE SUMMARY

- That, because of the known relationship of the calcium messenger system to growth, development, and cognitive function, new methods be developed to characterize disturbances in the calcium messenger system associated with lead exposure.
- That research be conducted on the effects of lead on affected organ systems (e.g., the reproductive system and the genitourinary system) and on the toxicokinetic behavior of lead (particularly bone lead) in human populations.
- That research be conducted to improve the understanding of mechanisms of low-dose lead toxicity, with emphasis on lead's effects on gene expression, calcium signaling, heme biosynthesis, and cellular energy production.
- That research be conducted to examine further the persistence and reversibility of lead's effects.

Exposure

The presence of mean blood lead concentrations in the U.S. population close to those which produce adverse health effects illustrates the importance of correctly measuring lead concentrations in sensitive populations. The committee recognized that exposure guidelines for health protection may be further reduced in the future.

The committee addressed measurement techniques for markers of both exposure and effect useful at low blood lead concentrations. Emphasis was placed on lead in physiologic media and EP, respectively. For the present and near future, the committee concluded that the primary screening tool to assess current lead exposure will be blood lead concentration, rather than EP concentration. At current body lead burdens of concern—i.e., those associated with the new CDC action level of 10 μg/dL—the EP technique is not sufficiently sensitive. (Evidence from diverse epidemiologic studies shows that the EP technique was not sufficiently sensitive even at the previous CDC guideline of 25 μg/dL.)

Current measurement techniques are capable of producing accurate and precise blood lead measurements. They include atomic-absorption spectrometry (AAS), anodic-stripping voltammetry (ASV), and thermal-ionization mass spectrometry (TIMS), all of which can be applied at parts-per-billion concentrations of lead in biologic media. Use of those methods assumes competence and strict attention to contamination control and other quality-assurance and quality-control (QA-QC) procedures. To achieve optimal methodologic utility and proficiency for routine and developmental needs, the committee particularly recommends

- That primary screening be done initially by measurement of lead in whole blood.
- That, given the current blood lead concentrations of concern, accurate and precise blood lead values be obtained, with strict attention to contamination control and other principles of QA-QC.
- That rigorous trace-metal clean techniques be established in sample collection, storage, and analysis.
- That a laboratory certification program be established, involving participation in external (blind) interlaboratory proficiency testing programs and analysis of lead in blood with concurrent analysis of appropriate reference materials.
- That, for more research-oriented purposes, standard reference materials be made available for such media as bone, blood, and urine, to allow laboratory evaluation of accuracy; this effort should be complemented by similar standards available for environmental media.
- That studies be conducted to explore the feasibility of applying ultraclean leadfree techniques to in vitro studies.
- That mass spectrometry with stable isotopes be used to investigate sources of environmental lead, as well as to examine lead metabolism in humans.

Sources and Accumulation

The committee identified the need for and acknowledged the rapidly developing availability of measurements for long-term lead accumulation during active exposure periods in sensitive populations, especially children and pregnant women. In so doing, it acknowledged that blood lead for routine purposes remains principally an index of recent exposure.

In vivo K- and L-line x-ray fluorescence measurements of long-term accumulated lead in trabecular and cortical bone of sensitive populations have been developed and evaluated in some detail, and they may be feasible as routine screening tools for selected sensitive populations in the future. The committee recommends

EXECUTIVE SUMMARY

- That more sensitive techniques for quantifying body burdens of lead in workers via bone lead dosimetry be developed.
- That, when radiation techniques are used for bone lead determinations, great care be taken that doses to individual subjects and populations, particularly the human conceptus, be carefully quantified according to National Council on Radiation Protection and Measurement (NCRP) guidelines.

The committee recognizes that the application of analytical techniques as described for the measurement of lead concentrations will require a large commitment of resources. Further, the establishment of new methods will require a significant investment of funds for research. However, the importance of the problem requires this commitment of manpower and funds.

Measuring Lead Exposure in Infants, Children, and Other Sensitive Populations

1
Introduction

Lead is a ubiquitous toxicant. It is especially toxic to young children and the fetus, and it was estimated that in 1984 about 6 million children and 400,000 fetuses in the United States were exposed to lead at concentrations to an extent that placed them at risk of adverse health effects (i.e., blood lead concentrations of at least 10 micrograms per deciliter (μg/dL) (CDC, 1991). Equally important, past screening programs were based on the blood lead concentrations of the U.S. Centers for Disease Control and Prevention (CDC) guideline of 25 μg/dL. Screening programs identify about 12,000 children with evidence of lead toxicity each year (ATSDR, 1988), but results of screening programs might seriously underestimate the magnitude of childhood lead exposure, primarily because few children are screened and because the false-negative rate of screening is high when screening is done with erythrocyte protoporphyrin.

In 1985, the CDC screening guideline for childhood blood lead associated with toxicity changed from a minimum of 30 μg/dL to a minimum of 25 μg/dL (CDC, 1985). The CDC guideline was set on the basis of health implications; of sensitivity, specificity, and cost-effectiveness of a screening program; and of the feasibility of effective intervention and followup. Evidence gathered since 1985 has shown that lead at less than half the previous CDC guideline (EPA, 1986a; Grant and Davis, 1989) can cause impairment of cognitive and physical development in children and increases in blood pressure in adults. The U.S. Environmental Protection Agency (EPA) Science Advisory Board has therefore proposed setting 10 μg/dL as a maximal safe blood lead con-

(EPA, 1990a). In response to the same evidence of effects at 10 µg/dL and even below, CDC has recently revised its 1985 statement (CDC, 1991). The 1991 statement's major points include (1) an acknowledgment that the current evidence on adverse effects associated with low-dose lead exposure requires a response from the federal medical and public-health community, (2) a reduction in the 1985 intervention or action concentration from 25 µg/dL to 10 µg/dL, and (3) the implementation of a multitiered, graded response that depends on measured blood lead concentrations. Responses will range from community-level actions to reduce lead exposure to emergency medical responses.

Persons exposed to lead and with blood lead above 10 µg/dL are likely to number in the millions, so appropriate methods for measuring concentrations of lead in blood at the revised guidelines will need to be developed and refined. In addition, new reliable and reproducible techniques for measuring lead in blood and other tissues, such as bone, will also need to be developed. Methods for detecting and measuring low-dose lead biologic markers are also needed, because the erythrocyte protoporphyrin (EP) test lacks sensitivity at blood lead concentrations below 25 µg/dL. For those reasons, the U.S. Agency for Toxic Substances and Disease Registry (ATSDR) requested that the National Research Council (NRC) provide information on techniques for measuring environmental exposure of sensitive populations to lead. The Board on Environmental Studies and Toxicology in the NRC Commission on Life Sciences formed the Committee on Measuring Lead Exposure in Critical Populations to meet the need.

PERSPECTIVE ON ISSUES

The finding of health effects of lead at blood concentrations previously considered low (i.e., about 10 µg/dL) is not surprising. First, the *typical* body burdens of lead in modern North Americans are at least 300 times greater than (reconstructed) burdens in North American Indians before European settlement (Ericson et al., 1991; Patterson et al., 1991). Thus, 10 µg/dL is not low, compared with the concentrations that used to prevail until relatively recently in humans. The increasing body burdens of lead with time are illustrated in Figure 1-1. Second, lead interferes with normal cellular calcium metabolism, causing intracellular

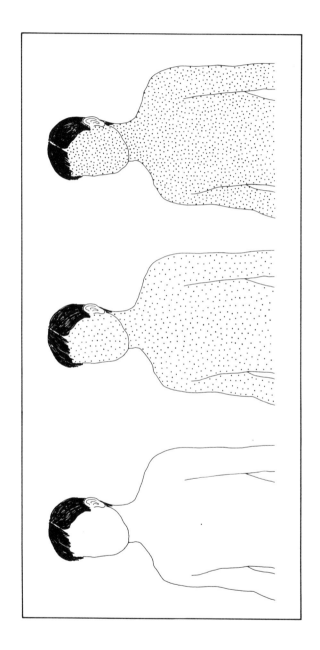

FIGURE 1-1 Body burdens of lead in ancient people uncontaminated by industrial lead (left); typical Americans (middle); people with overt clinical lead poisoning (right). Each dot represents 40µg of lead. Source: Patterson et al., 1991; adapted from NRC, 1980.

buildup of calcium. It binds normally to most calcium-activated proteins with 100,000 times the affinity of calcium; once bound, it interferes with the normal actions of these proteins.

Calcium and calcium-binding proteins serve as the messengers for many basic cellular processes. Some lead-caused disturbances, such as activation of protein kinease, show a dose-response relationship with no evidence of a threshold—hence the apparent absence of a threshold for some of the adverse health effects of lead. Third, death from encephalopathy or massive brain damage is common in children with untreated blood lead concentrations of 150 μg/dL and higher (NRC, 1972), and approximately 10% of the concentration that can cause death from brain damage might cause cognitive disturbances (as shown in epidemiologic studies).

Regulation of most toxic substances is based on safety factors: the presumed "safe" concentration for exposed people is set to be lower than the lowest-observed-effect concentration in humans by a factor of 10 to 100 (NRC, 1986). In contrast, the mean blood lead concentration of urban black children in 1978 was about one-sixth of the potentially fatal concentration. For all children, the mean was one-tenth of the potentially fatal concentration and was above the concentration at which decrements in IQ and other cognitive entities have since been established. The concentrations of lead in blood at which lead-abatement actions have been recommended over the past several years is shown in Figure 1-2. Cognitive effects of lead have been found in infants and children with blood lead concentrations of 10 μg/dL (ATSDR, 1988; Grant and Davis, 1989; Baghurst et al., 1992; Bellinger et al., 1992; Dietrich, 1992). Other studies have reported effects of lead at 10-15 μg/dL on growth rates, attained stature, birthweight, gestational age, auditory functioning, attention span, blood pressure, and some of the general metabolic pathways (Schwartz et al., 1986; Dietrich et al., 1987a,b, 1989; Schwartz and Otto, 1987, 1991; Shukla et al., 1987, 1991; Bellinger et al., 1988, 1991a; Bornschein et al., 1989; Thomson et al., 1989).

Prevention of disease is preferable to treatment, particularly if treatment is not certain to reverse damage. A wide variety of public agencies and offices are trying to reduce exposures to lead. The greatest successes have been in reducing lead in gasoline and food, and the introduction of new lead into paint and plumbing systems has been substantially reduced. But relatively little has been done to reduce exposure to

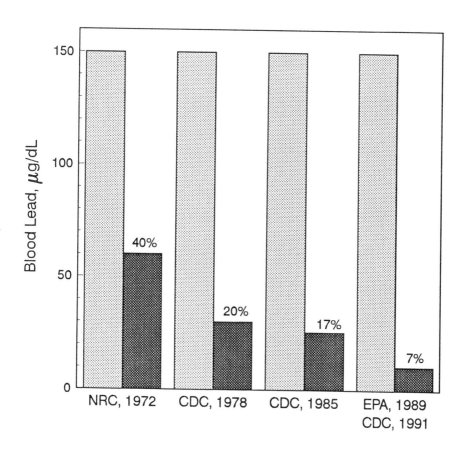

FIGURE 1-2 Children's blood lead concentrations at which lead-abatement action has been recommended. Cross-hatch columns, blood lead concentration that if untreated is potentially fatal to young child (150 µg/dL). Black columns, percentage of potentially fatal blood lead concentration at which intervention has been recommended.

existing lead in paint and plumbing systems and the reservoir of lead-contaminated soil and dust. For example, 42 million families live in housing that contains an estimated 3 million tons of lead in paints in the immediate environment (ATSDR, 1988)—equivalent to about 140 lb per household. Ingestion of as little as $2.4 \times 10^{-7}\%$ of that household burden each day (150 μg/day) would result in a steady-state aggregate lead concentration now considered toxic. Similarly, over 90% of U.S. housing units have lead-soldered plumbing (Levin, 1986). Occupational exposure at concentrations above those associated with reduced nerve conduction velocity, increased blood pressure, reduced reserve capacity for blood formation, and adverse reproductive effects is still common and legally permissible.

Figures 1-3 and 1-4 show the decline in gasoline lead use and the decline in food lead in the typical infant diet in the United States. The reductions have been accompanied by a substantial reduction in the average blood lead concentration of the population, as shown in Figure 1-3. However, children with high lead exposures to such sources as paint have not benefited proportionately, and the lack of substantial progress in reducing the source of their exposure constitutes a great failure in exposure prevention. Reducing lead in drinking water should be more easily accomplished by controlling corrosion. Apart from the benefits derived from reduced lead exposure, the economic saving in pipe and water-heater replacement would exceed the costs (Levin, 1986, 1987; Levin and Schock, 1991).

Once lead is mined and introduced into the environment, it persists. Over time, lead in various forms becomes available to the body as small particles. Most of the 300 million metric tons of lead ever produced (Figure 1-5) remains in the environment, largely in soil and dust. That explains, in part, why background concentrations of lead in modern North Americans are higher by a factor of 10^2-10^3 than they were in pre-Columbian Americans. Today's production evolves into tomorrow's background exposure, and despite reductions in the use of lead for gasoline, overall lead production continues to grow and federal agencies have not addressed the impact of future increases of lead in the environment.

Until very recently, lead poisoning had been perceived as a potentially fatal illness associated with acutely high exposure to lead and manifested as encephalopathy, acute abdominal colic, and acute kidney damage

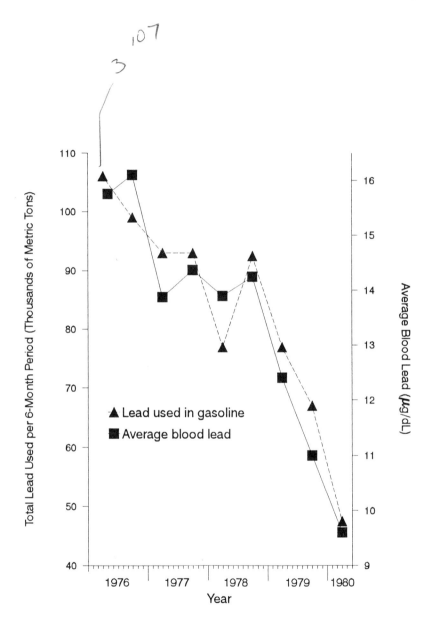

FIGURE 1-3 Lead used in gasoline production and average NHANES II blood lead (Feb. 1976-Feb. 1980). Source: Annest, 1983.

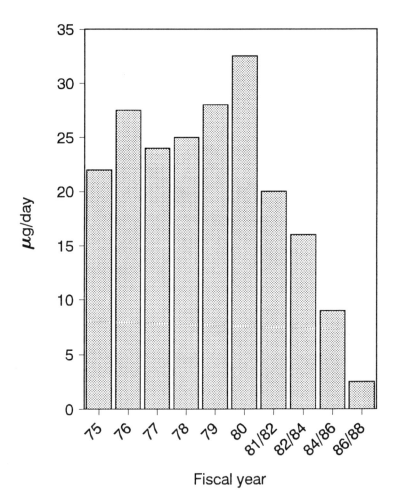

FIGURE 1-4   Average daily intakes of lead (based on FDA Total Diet Study) in infants 0-6 months old. Source: Capar, 1991.

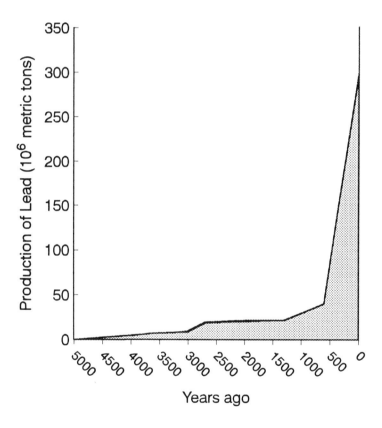

FIGURE 1-5 Cumulative production of lead over historic time. Source: Flegal and Smith, 1992. Reprinted with permission from *Environmental Research*; copyright 1992, Academic Press.

(Fanconi's syndrome). The association between chronic large exposure and peripheral neuropathy and gout, mainly in lead workers, has also been recognized. In the United States until about 1970, upper limits of acceptable or normal blood lead concentrations were set just below the concentrations associated with overt clinical illness (60 µg/dL in children, 80 µg/dL in adults), and "prevention" referred only to the prevention of clinical symptomatic episodes of lead poisoning. In the late 1960s and early 1970s, however, the concept of preventing lead toxicity became prevalent. In 1972, NRC published *Lead: Airborne Lead in Perspective*, the first comprehensive document on the subject published in the United States in almost 30 years. That report recommended a "further search for possible subtle effects of prolonged, low-level exposure to lead," inasmuch as the information available at the time did not "afford an adequate basis for the evaluation of this critical concern."

There has since been a great increase in both experimental and clinical investigations, and they have led to the conclusion that chronic low-level lead exposure, insufficient to produce recognizable clinical symptoms of toxicity, has adverse and probably long-lasting effects, particularly on neurodevelopment. An 11-year followup study has indicated that the neurodevelopmental effects persist and can have profound consequences for school achievement (Needleman et al., 1990). The neurodevelopmental studies have led to the identification of infants, children, and pregnant women (as surrogates for fetuses) at greatest risk of lead toxicity from low-level exposure. In this report, these are described as the most sensitive populations. The EPA Science Advisory Board and the Centers for Disease Control and Prevention concluded that published data clearly indicate that the upper limit of acceptable blood lead concentration is 10 µg/dL. This reduced acceptable blood concentration necessitates substantial improvement in analytic methods for measuring lead, as well as development of newer methods for measuring very low blood lead concentrations. Those measurement issues are the focus of this report.

HISTORICAL BACKGROUND

Lead was known and widely used by ancient civilizations. A lead statue displayed in the British Museum, discovered in Turkey, dates from 6500 BC (Hunter, 1978). Lead has been mined, smelted, and used in cosmetics, internal and topical medicinal preparations, paint pigments, and glazes since early in recorded history (Nriagu, 1983a). In ancient times, the uses of lead medicinally and as a food additive, sweetening agent, and wine stabilizer were probably the principal means by which humans became poisoned by lead. Food and beverage containers crafted from lead compounds, such as pewter, were also likely exposure routes. Since the Middle Ages, a particularly common means of exposure has been the adulteration of wine by lead (Stevenson, 1949; Wedeen, 1984).

Beginning of Public-Health Interest in Lead

Although it is probable that workers involved in the mining, smelting, and working of lead in ancient times were poisoned, technical details to support the assumption are not available. With the advent of the Industrial Revolution, however, interest in hazardous occupational lead exposures began to develop.

The clinical manifestations of lead poisoning were first described by Nikander, a Greek poet and physician, in the second century BC, who wrote:

> The harmful cerussa [white lead], that most noxious thing
> Which foams like the milk in the earliest spring.
> With rough force it falls and the pail beneath fills
> This fluid astringes and causes grave ills. . . .
> His feeble limbs droop and all motion is still.
> His strength is now spent and unless one soon aids
> The sick man descends to the Stygian shades.
> (Nikander, cited by Major, 1945, p. 312)

Physicians who came after Nikander also described the clinical manifestations of lead poisoning, but many failed to make a connection between the symptoms and the causative agent. Even today, the clinical diagnosis of lead poisoning from low doses is elusive (Harris and Elsea, 1967; Crosby, 1977; Wallace et al., 1985).

Today's interest in lead's impact on the health and occupational fields can be said to date from the 1839 publication of Tanquerel des Planches, in Paris, in which he described the clinical course of 1,207 persons with lead colic and the types of work that exposed them to lead. More than 800 of the cases were in painters or workers involved in the manufacture of white or red lead pigments.

Somewhat later, particularly during the industrialization of Europe, reproductive failures and congenital lead poisoning were described by Paul (1860). Workers recognized sterility, abortion, stillbirth, and premature delivery as common, not only among female lead workers, but also among the wives of men who worked in the lead trades (Oliver, 1911; Hamilton and Hardy, 1949; Lane, 1949). Indeed, those observations led a British Royal Commission in 1910 to recommend that women be excluded from the lead trades, a recommendation that was enforced in some countries by law (Lane, 1949). In the United States, occupational-hygiene actions toward protection of women from lead in the workplace have included the exclusion of women from lead exposure in an early stage of pregnancy (29 CFR 1910.1025) and even an effort to exclude all women of childbearing age. The U.S. Supreme Court has recently ruled that women of childbearing age cannot be excluded from workplaces with lead present.

Lead poisoning in children was first described by Gibson et al. in 1892 in Brisbane, Queensland, Australia. By 1904, Gibson, an ophthalmologist, had identified the source of lead and its probable route of entry into children, using both observation and experiment. He wrote:

I . . . advance a very strong plea for painted walls and railings as the source of the lead, and for the biting of finger nails and sucking of fingers, as in a majority of cases, the means of conveyance of lead to the patient.

In Europe, at about the same time, the general health hazards of lead in paint might have already been recognized. Awareness of the hazards was reflected in advertising of the period: Figure 1-6[1] depicts an adver-

---

FIGURE 1-6 Advertisement for Aspinall's enamel, which appeared in the Diamond Jubilee issue of the *Illustrated London News*, 1897. Note that it is not made with lead, is not toxic, and represents 60 years of progress. The latter probably refers to the work of Tanquerel des Planches in Paris in the 1830s. He published his famous treatise, "Les Maladies de Plumbe" in 1839. Note that Aspinall's enamel could be purchased in 1897 in Paris, London, and New York.

tisement appearing in England in 1897 emphasizing the nonpoisonous nature of the product, in contrast with toxic leaded paint.

Thomas and Blackfan in 1914 described the first American case of lead-paint poisoning (in a 5-year-old boy). They also offered the opinion that children must in some way be peculiarly susceptible to lead. On the occasion of the centennial of the Royal Children's Hospital in Brisbane in 1978, Fison (1978) cited Gibson's work and noted that at around the turn of the century physicians in southern Australia had been skeptical, because the condition seemed to stop abruptly at the Queensland border, and regarded the condition as a "delusion held by their despised colleagues in the primitive northern state." However, in the warm humid climate, paints weather quickly, and children would soon have their hands coated with powdery leaded material, which was inevitably carried to their mouths and digestive tracts.

After the report of Thomas and Blackfan in 1914, sporadic case reports appeared, and McKhann, in 1926, published a series of 17 cases of lead poisoning in children. That was followed by a classic report of lead poisoning due to the burning of battery casings in Baltimore homes in 1931 and 1932 (Williams et al., 1933). More case studies began to appear in the 1950s, as the condition became more widely known in larger cities in the United States, including New York, Chicago, Philadelphia, Boston, Cincinnati, St. Louis, and Cleveland. By 1970, the epidemiology of childhood lead poisoning was well established. Lin-Fu (1982) has summarized the history of childhood lead poisoning in the United States.

History of U.S. Childhood Lead-Screening Programs

Although several cases of childhood lead poisoning in the United States were reported in the first half of this century, little effort was made to understand the extent of poisoning in children until the 1950s, when caseworkers in a few large cities attempted to identify poisoned children as part of their family nutrition work. Limited results were obtained. In 1966, Chicago was the site of the first mass screening program where many poisoned children were identified; it was followed shortly by New York City and other large cities (Lin-Fu, 1980), where similar results were obtained. In 1971, the Surgeon General issued a statement that emphasized the need to shift the focus from identifying poisoned children to prevention.

The 1971 Lead-Based Paint Poisoning Prevention Act led to the Categorical Grants Program to help communities carry out screening and treatment programs. The act initiated a national effort to identify children with high blood lead concentrations and to attempt abatement of their environmental sources of lead. Funds appropriated under the act were first administered by the Bureau of Community Environmental Management of the Department of Health, Education, and Welfare and later by the Centers for Disease Control and Prevention. Annual expenditures under the act rose from $6.5 million in fiscal year 1972 to $11.25 million in fiscal year 1980. The money supported up to 62 screening programs in 25 states (NCEMCH, 1989).

While CDC conducted the categorical program, criteria for identifying lead toxicity underwent a number of changes. A confirmed blood lead concentration of 40 μg/dL had been used to "define" lead toxicity. The 1975 and the 1978 revisions of the earlier CDC guidelines (CDC, 1975, 1978) used 30 μg/dL or above and different definitions of high erythrocyte protoporphyrin (EP) concentrations to produce several risk categories. Over 2.7 million children were screened from July 1, 1972, to June 30, 1979; 183,452, or 7%, tested positive by the prevailing criteria. In 1981, over 500,000 children were screened; in 18,000, the results were "defined" as lead poisoning (CDC, 1982).

In 1981, the Maternal and Child Health Services Block Grant Act and the Omnibus Budget Reconciliation Act transferred the national administrative responsibility for childhood lead-poisoning prevention programs to the Division of Maternal and Child Health of the Bureau of Health Care Delivery and Assistance. Under the provisions of the block grant act, each state decides whether to use federal funds to support childhood lead-poisoning prevention efforts (NCEMCH, 1989). In the transition from the categorical to the block grant programs, the screening-data reporting requirement was eliminated. Furthermore, federal policy from the U.S. Office of Management and Budget discouraged continued reports of screening from existing programs.

In early 1985, CDC reduced its criterion of childhood lead poisoning (CDC, 1985). A blood lead concentration at or above 25 μg/dL in tandem with an EP concentration of 35 μg/dL or above was now considered evidence of potential lead toxicity. The 1985 statement made clear that adverse effects were recognized as occurring at blood lead concentrations below 25 µg/dL (so the chosen criteria represented the best compromise between health protection and practical matters related to limitations in screening methods at that time). The 1985 recommendations by CDC have now been re-evaluated, and an updated statement on lead-poisoning prevention was recently issued (CDC, 1991), which defines a lead concentration in whole blood of 10 µg/dL or greater as the action level, i.e., level of intervention.

The Lead Contamination Control Act of 1988 authorized, over 3 years, $20, $22, and $24 million per year for CDC to administer a childhood lead-poisoning prevention program. The money was to be given to state and local agencies to perform childhood lead screening for medical and environmental followup and education about lead poisoning. The act specifically stated that the money was not to supplant other funding for childhood lead-poisoning prevention. Although no money was appropriated for fiscal 1989, $4 million was appropriated for fiscal 1990, about $8 million for fiscal 1991, and $21.3 million for fiscal 1992. At present, national systematic collection of screening results does not exist.

The most recent national projection data on lead in children are available in the 1988 ATSDR report. The authors estimated that 250,000 children under the age of 6 had blood lead concentrations above 25 µg/dL in 1984. That estimate was based on data from NHANES II, census data, and projected environmental quantities of lead (ATSDR, 1988).

Sensitive Populations

People are not all equally vulnerable to the effects of environmental exposure to lead. That fact yields the idea of sensitive populations that should be protected by monitoring programs from the adverse health effects of lead exposure. The focus of this report is infants, children, and pregnant women (as surrogates for fetuses). These populations are the most sensitive to lead exposure and are defined by the committee as the sensitive populations for this report. Some vulnerability is intrinsic, such as the age-dependent vulnerability in developing organ systems, and some is extrinsic, such as that related to type or location of childhood exposure.

The magnitude of external exposure to lead varies among populations and influences both the severity of effects and the incidence of effects within a defined population. Besides the sensitive populations described in this report, other populations such as lead-industry workers may be at increased risk of adverse health effects because of large exposures. Nonetheless, the consideration of occupational exposures except where it may affect these sensitive populations is beyond the scope of this report. The time and magnitude of exposure may also influence the nature of lead intoxication. In addition, the constellation of organ systems affected can vary with age, with persistence of lead exposure, and with nutritional status, all of which are summarized in this report.

SCOPE AND ORGANIZATION OF THE COMMITTEE REPORT

The text and organization of the committee's report constitute a direct response to its charge, i.e., an assessment of appropriate, accurate, and precise methods for determining lead exposure in sensitive populations. The magnitude of human lead exposure deemed to be safe or without health effects continues to be reduced, to reflect improved identification of subtle toxic effects in sensitive populations. The committee considered measurement methods that could be useful in a world in which there might be no apparent threshold for particular neurobehavioral and other adverse health effects in vulnerable segments of the human population. The committee has evaluated exposure-measurement methods in the light of the existing broad range of population lead exposures.

This report has two main portions. One provides a summary of toxicity, public-health concerns and scientific context that helps to define the need for the committee's report. The second addresses the specifics of the committee's charge and some closely linked issues.

The background portion of this report consists of three chapters: this chapter has provided a perspective on key issues in the lead problem and a brief history of lead exposure assessment in the United States. Chapter 2 provides a summary of the toxicity of lead in sensitive populations. Chapter 3 deals with the nature and scope of source-specific environmental lead exposure in sensitive populations and the relative contributions of specific sources to adverse health effects.

The second portion consists of Chapters 4-6. Chapter 4 deals with the biologic basis of the markers of early effects of lead exposure. It summarizes the markers of exposure, effect, and susceptibility and establishes the biologic basis for the methods evaluated in Chapter 5. Chapter 5 describes the current and developing methods that are most suitable for lead exposure monitoring at low concentration, i.e., current CDC guidelines—including their population monitoring advantages and disadvantages. The methods are primarily those for use in internal (i.e., biologic) monitoring of lead exposure. They are described in relation to sampling and transporting, instrumentation, quality assurance and quality control, and statistical methods. The chapter describes the public-health implications of monitoring sensitive populations for lead exposure at low concentrations as recommended by CDC. Chapter 6 sets forth a comprehensive set of recommendations dealing with both the specifics and the generalities of the committee's charge.

2
Adverse Health Effects of Exposure to Lead

Exposure to lead produces a variety of adverse health effects in sensitive populations through its impact on different organs and systems. The nature of the effects is a complex function of such factors as the magnitude of exposure, the physiologic and behavioral characteristics of the exposed person, and the relative importance of the lead-injured organ or system to overall health and well-being. The toxic effects of lead range from recently revealed subtle, subclinical responses to overt serious intoxication. It is the array of chronic effects of low-dose exposure that is of current public-health concern and that is the subject of this chapter. Overt, clinical poisoning still occurs, however, and is also discussed here. We have several reasons for emphasizing low-dose exposure. As recently noted by Landrigan (1989), the subtle effects of lead are bona fide impairments, not just inconsequential physiologic perturbations or slight decreases in reserve capacity. And the effects are associated with magnitudes of lead exposure that are encountered by a sizable fraction of the population in developed countries and thus are potentially found in very large numbers of people.

This chapter summarizes key points about the health effects of exposure of sensitive populations to lead. It deals specifically with various adverse effects of lead in sensitive populations (with emphasis on effects of low-dose exposure), persistence of some important health effects, molecular mechanisms of lead toxicity, and dose-effect relations. As described below, clinical lead poisoning differs between children and adults, in part because their organ systems are affected in different ways and to different extents. In addition, some people are more vulnerable to lead toxicity and have increased sensitivity, because they suffer from disease, lack proper nutrition, or lack adequate health care. These factors influence exposure patterns and the biokinetics of lead absorption.

CLINICAL INTOXICATION IN CHILDREN

Childhood lead poisoning involves injury in at least three organ systems: the central nervous system (specifically, the brain), the kidney, and the blood-forming organs. Other systems are also affected, but the nature of their toxic injury has not been as well characterized.

Central Nervous System Effects

The nature of lead-associated overt central nervous system injury in children differs with the degree of lead exposure. Blood lead concentrations of about 100-150 μg/dL are associated with a high probability of fulminant lead encephalopathy. Before the widespread clinical use of chelation therapy, lead encephalopathy carried a high rate of mortality, about 65% (Foreman, 1963; NRC, 1972). The use of chelation therapy, pioneered by Chisolm and co-workers (NRC, 1972), has reduced mortality to 1 or 2%, or even less, if the poisoning is recognized and dealt with. The range of blood lead concentrations reported in association with encephalopathy is quite large (NRC, 1972), owing to such factors as individual differences in toxicokinetics and in timing of lead measurement and treatment.

Children are much more sensitive than adults to the neuropathic effects of lead. The central nervous system is principally involved in children and the peripheral nervous system in adults (Chisolm and Harrison, 1956; NRC, 1972; Chisolm and Barltrop, 1979; Piomelli et al., 1984), and thresholds of blood lead concentration for neurofunctional measures are lower in children. The common neuropathologic findings in fatal childhood lead encephalopathy (Pentschew, 1965) are cerebral edema, structural derangement in capillaries, neuronal necrosis, and neuronal loss in isocortex and basal ganglia.

Many children who survive an episode of lead encephalopathy have permanent neurologic sequelae, including retardation and severe behavioral disorders (Byers and Lord, 1943; Perlstein and Attala, 1966; Rummo et al., 1979).

Renal Effects

Kidney injury in childhood plumbism is most often seen in overt poisoning and involves damage to the proximal tubule. In poisoning with encephalopathy, Chisolm (1962, 1968) has found the presence of the full, albeit transitory, Fanconi syndrome: glycosuria, aminoaciduria, hyperphosphaturia (with hypophosphatemia), and rickets. In overt toxicity after a range of exposures, aminoaciduria is the most consistent finding (Chisolm, 1962).

In chronic high-dose lead exposure, aminoaciduria appears to be the most consistent nephropathic finding. In a group of children with blood lead concentrations of 40-120 µg/dL, Pueschel et al. (1972) found aminoaciduria in those with blood lead of 50 µg/dL or more.

The role of childhood lead poisoning as a known contributor to early adult chronic nephritis found in Australia (e.g., Henderson, 1954; Emmerson, 1963) has not been identified in the United States (Tepper, 1963; Chisolm et al., 1976).

Hematologic Effects

Anemia is common in severe chronic lead poisoning and is reported to be associated with blood lead concentrations of 70 µg/dL and higher (NRC, 1972; CDC, 1978; Chisolm and Barltrop, 1979). However, reanalyses of the hematologic data of Landrigan et al. (1976) in children by Schwartz et al. (1990) indicated that anemia as indexed by hematocrit is present with blood lead concentrations below 70 µg/dL. Typically, the anemia is mildly hypochromic and normocytic and arises from a combination of reduced hemoglobin formation (resulting from either impaired heme synthesis or globin chain formation) and reduction in erythrocyte survival because of hemolysis (Waldron, 1966; Valentine and Paglia, 1980).

INTOXICATION IN ADULTS

The most extensive adult studies are of workers occupationally exposed to lead in battery recycling, lead smelting, alkyl lead manufacturing, plumbing, and pipefitting. These studies are described in other reports (EPA, 1986a; ATSDR, 1988) of lead exposure and are not the focus of this report. This report examines principally the effects of lead exposure in pregnant women as a sensitive population. Other adult populations may be at increased risk of lead intoxication because of large exposures, but they are beyond the scope of this report. For this reason, only a brief summary of effects in adults is presented below.

Central Nervous System and Other Neuropathic Effects

Although lead poisoning after very large exposures in adults can produce central nervous system injury, the exposure threshold is much higher in adults than in children. Blood lead concentrations associated with adult encephalopathy are well above 120-150 μg/dL. The features of adult lead encephalopathy, which can be as abrupt in onset in adults as in children, have been described by Aub et al. (1926), Cantarow and Trumper (1944), and Cumings (1959). They include dullness, irritability, headaches, and hallucination, progressing to convulsions, paralysis, and even death.

The more typical neuropathologic outcome of adult lead poisoning is peripheral polyneuritis involving sensory or motor nerves. There is often pronounced motor dysfunction, such as wrist drop and foot drop in the more advanced cases (Feldman et al., 1977). Changes include segmental demyelination and axonal degeneration (Fullerton, 1966), often with concomitant endoneural edema of Schwann cells (Windebank and Dyck, 1981).

Renal Effects

Occupational chronic lead nephropathy, the most important category of lead-associated kidney injury in adult populations, has been heavily studied for many years in Europe, but not as well in the United States.

In the studies of Wedeen et al. (1975, 1979), renal dysfunction has been established in U.S. lead workers, many of whom had no history of prior lead poisoning.

Generally, the Fanconi syndrome of acute childhood poisoning is not seen in adults with chronic lead poisoning. Proximal tubular injury from lead in adults at early stages of nephropathy is difficult to detect in workers, because of extensive renal reserve (Landrigan et al., 1982). Hyperuricemia is frequent probably because of increased uric acid production (Granick et al., 1978).

Lead has been clearly demonstrated to produce tubular nephrotoxicity in humans and rodents after acute or chronic exposure (Goyer and Rhyne, 1973; Wedeen et al., 1986; Ritz et al., 1988). Tubular proteinuria is a well-known manifestation of metal nephrotoxicity, but inconsistently reported in lead nephrotoxicity (Bonucci and Silvestrini, 1989; Goyer, 1989), perhaps because of the lack of sensitive and specific protein assays (Bernard and Becker, 1988). With the advent of two-dimensional gel electrophoresis (O'Farrell, 1975) and highly sensitive silver staining methods (Merril et al., 1981), it should be possible to separate various nonreabsorbed proteins from the urinary filtrate into lead-specific patterns at an early stage of tubular injury or monitor the low-molecular-weight proteins, such as retinol-binding protein, which is stable at the pH of normal urine (Bernard and Becker, 1988).

Hematologic Effects

Lead workers often show evidence of both marked impairment of heme biosynthesis and increased erythrocyte destruction (EPA, 1986a). Characteristic biochemical and functional indexes of those impairments include increased urinary delta-aminolevulinic acid and erythrocyte zinc protoporphyrin, increased cell fragility, and decreased osmotic resistance, which combine to produce anemia (Baker et al., 1979).

REPRODUCTIVE AND DEVELOPMENTAL EFFECTS

Reproductive and Early Developmental Toxicity

Reproductive toxicity resulting from the high-dose lead exposure is well established (Rom, 1976). Much of the early literature focused on an increased incidence of spontaneous abortion and stillbirth associated with lead exposure in the workplace (Paul, 1860; Legge, 1901; Oliver, 1911; Lane, 1949). In addition, lead was used as an abortifacient in England (Hall and Cantab, 1905). These outcomes, which are far less common today, presumably involve some combination of gametotoxic, embryotoxic, fetotoxic, and teratogenic effects and define the upper end of the spectrum of reproductive toxicity in humans. Since these earlier reports, industrial exposure of women of childbearing age was restricted by improved industrial hygiene practices, but a recent U.S. Supreme Court decision ruled exclusion illegal. The decision was based on the premise of equal access to the workplace, not on insufficiency of evidence of toxic harm.

Epidemiologic studies of exposed women have reported reproductive effects of lead exposure in both nonoccupational groups (Fahim et al., 1976; Nordstrom et al., 1978a,b) and occupational groups (Panova, 1972). Deficiencies in the design of the studies prevent definitive conclusions, but the studies have helped to direct attention to a potential problem.

Very early preimplantation loss can easily go undetected and might be occurring after moderate-dose and perhaps even low-dose exposure. With the advent of human chorionic gonadotropin assays, it is now possible to detect the onset of pregnancy and early fetal loss during the first 1-2 weeks of pregnancy. Savitz et al. (1989) used data from the National Natality and Fetal Mortality Survey, a probability sample of live births and fetal deaths to married women in 1980, to show that maternal employment in the lead industry was a risk factor for negative pregnancy outcomes, including stillbirth (OR = 1.6) and preterm birth (OR = 2.3). No systematic study has been conducted of the effects of increased lead stores on early fetal loss in women who may have incurred substantial lead exposures during their childhood or during a prior period of employment in a lead-related trade. Such studies are warranted, given the known reproductive toxicity of large exposure to lead.

Several prospective studies have examined the issue of lead's involvement in spontaneous abortion, stillbirth, preterm delivery, and low birthweight. Women in the studies in Boston (Bellinger et al., 1991b), Cleveland (Ernhart et al., 1986), Cincinnati (Bornschein et al., 1989), and Port Pirie (McMichael et al., 1986; Baghurst et al., 1987a) had average blood lead concentrations during pregnancy of 5-10 µg/dL; almost all had blood lead concentrations less than 25 µg/dL. The Glasgow (Moore et al., 1982) and Titova Mitrovica (Graziano et al., 1989; Murphy et al., 1990) cohorts had average blood lead concentrations of about 20 µg/dL. None of those studies reported an association between maternal blood lead concentrations and spontaneous abortion or stillbirth. However, the Cincinnati and Port Pirie studies found a lead-related decrease in duration of pregnancy, and the Glasgow, Cincinnati, and Boston studies reported a lead-related decrease in birthweight. The Boston study found an increased risk of intrauterine growth retardation, low birthweight, and small-for-gestational-age deliveries at cord blood lead concentrations of 15 µg/dL or more. The Port Pirie study found that the relative risk of preterm delivery increased 2.8-fold for every 10-µg/dL increase in maternal blood lead. In the Cincinnati study, gestational age was reduced about 0.6 weeks for each natural log unit increase in blood lead, or about 1.8 weeks over the entire range of observed blood concentrations. Even after adjustment for the reduced length of pregnancy, the Cincinnati study found reduced infant birthweight (by about 300 g) and birthlength (by about 2.5 cm), and the Port Pirie group reported reduced head circumference (by about 0.3 cm) (Baghurst et al., 1987b). Findings from some of the prospective studies have been extensively reviewed (Davis and Svendsgaard, 1987; Ernhart et al., 1989; Grant and Davis, 1989). However, some striking inconsistencies, yet to be explained, characterize the data on the relationship between prenatal lead exposure and fetal growth and maturation. For instance, in the large cohort (N = 907) of women residing in Kosovo (Factor-Litvak et al., 1991), no associations were seen between midpregnancy blood lead concentrations (ranging up to approximately 55 µg/dL) and either infant birthweight or length of gestation.

Several studies have also looked for evidence of teratogenicity (Needleman et al., 1984; Ernhart et al., 1986; McMichael et al., 1986). Needleman et al. (1984), in a retrospective study of the association between cord blood lead and major or minor malformations in a cohort of 4,354 infants, found a significant increase in the number of minor anomalies observed per child, but no malformation was found to be associated with lead. Unexpectedly, several other factors, such as premature labor and neonatal respiratory distress, were found to be reduced with increased blood lead. Both Ernhart et al. (1986) and McMichael et al. (1986) tried but failed to replicate these findings; however, these studies lacked the power to detect the small effects reported by Needleman et al. (1984). The Needleman et al. study is important because the minor anomalies in question might reflect general fetal stress and predict developmental disorders (Marden et al., 1964).

Evidence is accumulating that relatively small increases in maternal blood lead during pregnancy can be associated with delayed or retarded growth. Shukla et al. (1987) reported that 260 infants from the Cincinnati prospective lead study experienced retardation in covariate-adjusted growth. More specifically, they found that infants born to women with lead concentrations greater than 8 μg/dL during pregnancy grew at a lower than expected rate if increased lead exposure continued during the first 15 months of life. Conversely, if postnatal lead exposure was small, the infants grew at a higher than expected rate; that suggests a catchup in growth after fetal growth suppression. No lead-related growth effects were observed in infants born to women with blood lead concentrations less than 8 μg/dL. In a later analysis of stature at 33 months of age, Shukla et al. (1991) reported that sustained increases in lead exposure above 20 μg/dL throughout the first 33 months of life are associated with reduced stature. However, prenatal exposure was no longer related to stature at 33 months of age. The reported indication of fetal toxicity is consistent with other previously discussed markers of lead-related fetal toxicity. It is also consistent with cross-sectional studies of lead's relation with physical size.

Several points emerge from a review of those studies, apart from a lead-related retardation of growth itself. First, the specific manifestations of the fetal insult vary among cohorts and might reflect lead's interaction with such cofactors as adequacy of prenatal care, maternal age, ethnicity, and nutritional status. Second, the blood lead concentrations associated with adverse fetal development are low (10-15 μg/dL or even lower) and comparable with those found in a substantial fraction of women of childbearing age (ATSDR, 1988). The validity of the reported association between fetal lead exposure and markers of adverse fetal development is strengthened by the observed negative association between maternal or fetal blood lead concentrations and early physical growth and cognitive development (Bellinger et al., 1987; Dietrich et al., 1987a,b; Vimpani et al., 1989). Thus, the birth-outcome measures, early physical-growth measures, and early measures of infant development can be viewed as potentially reflecting the fetal toxicity of lead.

Gametotoxicity of lead has been studied primarily in male lead workers. Lancranjan et al. (1975) noted lead-associated disturbances of reproductive competence in lead workers; blood lead concentrations of about 40 µg/dL were associated with asthenospermia and hypospermia, and higher concentrations with teratospermia. Erectile dysfunction was observed in the lead workers, but did not seem dose-dependent. Zielhuis and Wibowo (1976) criticized the design and results of that study, noting potential underestimation of blood lead concentrations.

Wildt et al. (1983) noted that lead-battery workers with blood lead concentrations over 50 µg/dL showed prostatic and seminal vesicular dysfunction compared with controls. However, their study had a number of methodologic problems concerning the measures of dysfunction and exposure monitoring (EPA, 1986a).

More recently, Assennato and co-workers (1986) reported sperm count suppression in lead-battery workers in the absence of endocrine dysfunction. Rodamilans et al. (1988) found that duration of lead exposure of smelter workers was variably associated with endocrine testicular function: workers who had been employed for more than 3 years had decreases in serum testosterone, steroid-binding globulin, and free-testosterone index. In both studies, the mean blood lead concentrations were over 60 µg/dL.

We have already noted longitudinal studies of lead's effects on growth and development in young children. Cross-sectional data are also available from a large population survey. Schwartz et al. (1986) reported that postnatal exposure of U.S. children affects later growth, according to analysis of the large NHANES II data set with respect to height, weight, and chest circumference as a function of blood lead concentration. The three growth milestones in children under 7 years old were significantly and inversely associated with blood lead concentration: height, $p < 0.0001$; weight, $p < 0.001$; and chest circumference, $p < 0.026$. The association was present over the blood lead concentration range of 5-35 µg/dL. These results are consistent with those of Frisancho and Ryan (1991), who found an inverse association between blood lead level and stature in a cohort of 1,454 5-12 year old children in the Hispanic HANES data set, and those of Lauwers and co-workers (1986) in Belgium, who noted statistically significant and inverse associations among growth indexes and blood lead concentration in children up to the age of 8 years. Nonquantitatively, reduced stature has been seen in children chronically exposed to lead (Johnson and Tenuta, 1979).

Angle and Kuntzelman (1989) in a retrospective pilot study examined 30 children with increased blood lead concentrations (over 30 μg/dL) and erythrocyte protoporphyrin relative to those in a control group. Growth velocity, higher in the high-lead group before 24 months of age, reverted to a net retardation after this age, compared with values in controls. In a longitudinal followup study (Markowitz and Rosen, 1990), lead-poisoned children showed reduced growth velocity, compared with that in age-matched control subjects. Furthermore, impaired growth velocities in the lead-poisoned children did not change substantially from baseline after chelation therapy.

The data on children suggest that endocrinologic disturbances can occur at sensitive points in anthropometric development. Endocrine dysfunction in lead workers with relatively high lead exposure is known (Sandstead et al., 1970; Robins et al., 1983).

Huseman and co-workers (1987) found that height in two lead-poisoned children dropped to below the tenth percentile during intoxication; both subjects demonstrated depressed thyroid-stimulating hormone (TSH) responses to thyrotropin-releasing hormone (TRH), and one showed depression in resting TSH concentrations.

Cognitive and Other Neurobehavioral Effects

Information about the effects of low-level exposure to lead has been obtained principally from two types of epidemiologic studies. One is the cross-sectional or retrospective cohort study, in which children's lead exposure and development are assessed at the same time or in which past lead exposure is estimated. The second type is the prospective longitudinal study, in which children's exposure and development are assessed on multiple occasions. Each type of study has strengths and weaknesses. For clarity, the findings from each type are discussed separately.

Prospective Longitudinal Studies

The findings pertaining to the association between indices of prenatal lead exposure and early development are mixed. In some cohorts, prenatal exposures corresponding to maternal or cord blood lead concentrations of 10-20 μg/dL were associated with early developmental delays. In the Boston cohort, infants with cord blood lead concentrations between 10 and 25 μg/dL manifested a performance deficit of 4-8 points between 6 and 24 months of age, relative to infants with cord blood lead concentrations below 3 μg/dL (Bellinger et al., 1984a; 1986a,b; 1987). In the Cincinnati cohort, developmental scores at 3 and 6 months of age declined by 6-7 points for each increase of 10 μg/dL in prenatal lead concentrations in the range of 1-27 μg/dL (Dietrich et al., 1987a,b); in addition, 12-month Mental Development Index (MDI) scores were inversely related to infants' blood lead concentrations at 10 days of age. In the Cleveland cohort, increased cord blood lead concentrations were significantly associated with increased numbers of neurologic signs, and increased maternal blood lead concentrations with lower scores on the Bayley Scales and the Kent Infant Development Scale at age 6 months (Ernhart et al., 1986, 1987).

Quite different results were reported from the Australian studies. In the Port Pirie study, developmental assessments were first administered at 2 years of age, at which time MDI scores from the Bayley Scales were not associated with average antenatal, maternal, or cord blood lead concentrations (Baghurst et al., 1987b; Wigg et al., 1988). In the Sydney cohort, neither maternal nor cord blood lead concentration was inversely related to any index of children's development at 6, 12, 24, 36, or 48 months (Cooney et al., 1989a,b). In fact, cord blood lead concentration was positively associated with infants' motor development even after adjustment for covariates. Exposure misclassification is a potential problem in this study. At 12, 18, and 24 months of age, half the children provided capillary (fingerstick) blood samples, and the other half venous blood samples. Given the potential difficulties associated with capillary samples, the mixing of sampling methods at several ages complicates the effort to establish the relative exposures of the children in the cohort. For example, the Sydney group found that at age 3 the average lead concentration in capillary samples was 30% greater than that in venous samples. Mahaffey et al. (1979) noted a similar positive bias of 20% for capillary versus venous samples in the NHANES II data. Although the impact of those differences on exposure assessment in the Sydney cohort is uncertain, the investigators' concern over contamination of the early capillary samples prompted the recruitment of an additional 123 children.

The association between prenatal or perinatal exposure and indexes of overall development was apparent beyond the first year in the Boston cohort (Bellinger et al., 1987) and to a limited extent in the Cincinnati cohort (Dietrich et al., 1989), but not in the Cleveland cohort (Ernhart et al., 1987). In the Boston study, the association between prenatal exposure and children's performance attenuated after 2 years of age. However, children who had high prenatal exposure and high exposure at age 57 months (blood lead concentration greater than 10 μg/dL) "recovered" to a smaller extent than did children with high prenatal exposures but lower exposures at age 57 months (Bellinger et al., 1991a). In the Cincinnati cohort, neonatal blood lead concentration (measured at 10 days of age) was inversely associated with performance at age 4 years on all subscales of the Kaufman Assessment Battery for Children (K-ABC), but only among the more disadvantaged children (Dietrich et al., 1991). This association was not evident at 5 years of age on the K-ABC (Dietrich et al., 1992) or at 6.5 years on the Wechsler Intelligence Scale for Children-Revised (Dietrich et al., 1993a). Neonatal blood lead levels were, however, inversely associated with fine motor performance on the Bruininks-Oseretsky Test of Motor Proficiency (Dietrich et al., 1993b).

Several factors could account for the inconsistencies in findings across studies. First, infants might generally be able to compensate for early adversities associated with lead exposure. Second, the adverse impact of the substantial postnatal rise in the exposures of the children in most cohorts might have obscured a persisting effect of prenatal exposure on development. In the Port Pirie study, the absence of an association between antenatal or cord blood lead concentrations and 2-year Bayley scores could reflect the attenuation of an association that would have been detected if assessments had been carried out before age 2. Third, the impact of competing risks for poor development among disadvantaged infants might have overwhelmed a persisting but small effect of prenatal lead exposure. Fourth, the expression of lead insult might be modified over time by the child's social environment. Although an association between prenatal exposure and indexes of overall development might not persist, associations could emerge with respect to more specific aspects of development. For instance, in the Cincinnati cohort, the association between prenatal exposure (maternal blood lead concentration during pregnancy) and performance on the Bayley Scales attenuated by the time the children were 2 years old, but an inverse association was found between prenatal exposure and children's scores on the Fluharty Speech and Language Screening Test (Dietrich, 1991) at age 30 months. Fifth, the loss in power resulting from cohort attrition might have reduced the probability that a persisting association would be detected.

The latter hypothesis is contradicted, however, by a pattern of increasing consistency in data from the various studies supporting an inverse association between blood lead levels measured in the postnatal period and cognitive function in the late preschool and school-age period. In the Boston cohort, blood lead concentration across a range of 3-20 μg/dL at age 2 years was associated with a decrease of approximately 6 points in children's General Cognitive Index (GCI) scores on the McCarthy Scales at age 57 months (Bellinger et al., 1991a). The coefficients associated with other postnatal blood lead measurements as well as with dentin concentrations were also negative but not statistically significant. The inverse association between cognition and blood lead was still apparent at age 10 years. Children's IQ scores on the Wechsler Intelligence Scale for Children-Revised declined approximately 6 points for each rise of 10 μg/dL in blood lead level at age 2 years, while scores on the Kaufman Test of Educational Achievement declined approximately 9 points (Bellinger et al., 1992). In this cohort, the mean blood lead level at age 2 years was less than 7 μg/dL, with 90% of the values below 14 μg/dL. In the Port Pirie cohort, an increase from 10 to 31 μg/dL in a cumulative index of postnatal blood lead concentrations (particularly concentrations up to age 4 years) was associated with a decrease of approximately 7 points in GCI scores at age 4 years (McMichael et al., 1988) and a decrease of 4-5 points in WISC-R IQ scores at age 7 years (Baghurst et al., 1992). In the data for both studies, no threshold is discernable for the association between increased blood lead level and decreased performance. Moreover, in both studies, children's scores on the Perceptual-Performance subscale of the McCarthy Scales (and the Memory Scale as well in the Port Pirie study) and WISC-R Verbal IQ were most strongly associated with postnatal lead exposure. In the Cincinnati cohort, later postnatal blood lead concentrations, as well as indexes of lifetime blood lead, were weakly associated with children's scores at 4 and 5 years of age on the Simultaneous Processing subscale of the K-ABC, which, like the Perceptual-Performance subscale of the McCarthy Scales, assesses primarily visual-spatial and visual-motor skills

(Dietrich et al., 1991, 1992). Some of these associations were not statistically significant after "full" covariate adjustment, however. Right ear auditory processing skills at age 5 years, assessed by the Filtered Word subtest of the Screening Test for Auditory Processing Disorders, were significantly associated with postnatal blood lead concentrations as well (Dietrich et al., 1992). Scores on the Auditory Figure-Ground subtest were not associated with lead exposure. Assessments of this cohort at age 6 indicated that high postnatal blood lead levels (especially those measured around ages 4 and 5 years) were significantly associated with lower scores on WISC-R IQ (Performance IQ only) (Dietrich et al., 1993a), and various indices of both gross-motor and fine-motor function (Dietrich et al., 1993b).

Preliminary findings from the Yugoslavian prospective study indicate a significant inverse association between blood lead concentration at 2 years of age and concurrent MDI scores, corresponding to a decrease of 2.9 points as blood lead increased from 10 to 30 μg/dL (Wasserman et al., 1991).

The Cleveland study has provided little evidence that postnatal lead exposure is associated with children's development (Ernhart et al., 1989). A significant inverse association between blood lead concentration at age 2 years and IQ scores at age 4 was found, however, if four children identified as influential by regression diagnostics were excluded. The investigators discounted this finding on statistical grounds, however. In the Sydney study, a composite exposure index consisting of blood lead concentrations measured during the first year of life was weakly associated ($p = 0.07$) with children's adjusted GCI scores on the McCarthy Scales. It appears, however, that the association was positive, rather than negative; i.e., children with greater exposures achieved higher scores (Cooney et al., 1989a,b).

Inconsistencies in the data preclude drawing inferences about modifiers of any association between lead and development. Among children 6 months old and 4 years old in the Cincinnati cohort and children 18 and 24 months old in the Boston cohort, the inverse association between neonatal blood lead concentration and MDI scores was stronger among children below the median social class (Dietrich et al., 1987a,b, 1991; Bellinger et al., 1988). Such interactions have not been observed in all studies, however, or even at other ages within the Boston and Cincinnati cohorts. In addition, the performance of 6-month-old boys in the Cincinnati cohort was more strongly associated with blood lead concentration than was the performance of 6-month-old girls. To judge by estimates of performance change between 24 and 57 months of age in the Boston cohort, boys recovered more slowly than girls from the adverse effects of higher prenatal exposure. That is consistent with substantial evidence that a wide range of developmental neuropsychiatric disorders are more prevalent among boys than girls (Gualtieri, 1987). In the Port Pirie cohort, however, at ages 2, 4, and 7 years, the performance decrement of girls has consistently been found to be greater than that of boys (McMichael et al., 1992; Baghurst et al., 1992).

It is clear that there are points of both agreement and disagreement in the findings of the prospective studies. A variety of methodologic and substantive explanations can be posited and, at this juncture, it is not yet clear which are correct. In terms of methodologic factors, false positive findings (Type I errors) due to multiple comparisons or to incomplete adjustment for confounding could be responsible for the associations observed in some studies between lead exposure and cognitive development. False negative findings (Type II errors) due to factors such as statistical "over-control" or exposure misclassification may be responsible for the lack of associations reported in some studies.

In terms of substantive explanations, it is possible that the strength of the association, or the likelihood of detecting an association, depends on population characteristics that are not comparable in the various cohorts (e.g., socioeconomic status, medical risk status, lead exposure profiles). One would not necessarily expect all studies to produce the same results, and, indeed, they have not (Mushak, in press). When findings differ, however, the information yield is likely to be the greatest. Each study is likely to contribute only part of the answer to the general question, "Under what exposure conditions do different populations of children manifest a lead-associated impact on growth and development?" It is clear that the complete answer to this question is unlikely to be simply "all" or "none."

In summary, there is relatively little consistency across the set of prospective studies in terms of the association between indices of prenatal lead exposure and later cognitive function. In contrast, as the length of follow-up has increased to include assessments at school-age, striking consistencies are emerging, with all 3 studies (Boston, Cincinnati, Port Pirie) reporting significant inverse associations between blood lead levels measured in the first few postnatal years and intellectual performance at ages 6 to 10 years.

Table 2-1 summarizes the findings from prospective studies to date with respect to reproductive outcome and early cognitive development. An additional prospective study is being conducted in Mexico City (Rothenberg et al., 1989) but follow-up data at ages older than 30 days have not yet been published.

Cross-sectional and Retrospective Studies

Most of the recent cross-sectional studies of lead and children's cognition have been reviewed by Grant and Davis (1989). Here, an effort is made to identify major themes, including issues on which the data are inconsistent. For reference, basic features of the major studies are listed in Table 2-2.

Because each study has included a global assessment of children's intelligence, this outcome provides the strongest basis for interstudy comparison. To assess the degree to which the results of various studies support a common dose-response relationship between lead exposure and IQ, the mean IQ scores of children in different exposure groups from the various studies are plotted together in Figures 2-1 and 2-2. For studies that report a partial regression coefficient as the measure of association, the best fit line is presented if the authors also provide the intercept or if it can be discerned from a figure in the original report. Information from such studies can contribute to the effort to identify a lowest-observed-effect concentration, if adequate steps are taken to assess the underlying assumption that the lead-IQ association is linear over the range of exposures represented in a cohort. Integrating the findings from separate studies would be facilitated if, in reporting the association between lead and IQ (or any other outcome), investigators provided quantitative measures of effect size, such as regression coefficients and standard errors or adjusted means and standard errors, rather than simply p values, correlations, or percentages of variance (or incremental variance) attributable to lead. Because of differences in the measurement and interpretation of blood lead and tooth lead, the results of studies relying on these exposure indexes are plotted separately. The nature and extent of adjustment for confounding varies considerably from study to

TABLE 2-1 Prospective Studies

Study Site: Key References	No.[a]	Blood Lead Assessment	Outcome Assessment[b]	Major Findings
Boston: Bellinger et al., 1984a; 1986a,b; 1987; 1988; 1989b; 1990; 1991; 1991a; 1992	249	Cord low: <3 µg/dL medium: 6-7 high: 10-25 6 mo: $\bar{x} = 6.2$, SD = 7.1 12 mo: $\bar{x} = 7.7$, SD = 6.5 18 mo: $\bar{x} = 7.6$, SD = 5.7 24 mo: $\bar{x} = 6.8$, SD = 6.3 57 mo: $\bar{x} = 6.4$, SD = 4.1 10 yr: $\bar{x} = 2.9$, SD = 2.4	6 mo: BSID 12 mo: BSID 18 mo: BSID 24 mo: BSID 57 mo: MSCA 10 yr: WISC-R K-TEA	1. Mental Development Index of BSID inversely related to cord-blood lead group at all ages between 6 and 24 mo of age. 2. The inverse associations strongest for children below median social class. 3. Mental Development Index scores not related to blood lead concentrations measured in first 2 yr of life. 4. General Cognitive Index scores at 57 mo inversely related to blood lead concentration measured at 24 mo of age. 5. General Cognitive Index scores not associated with cord-blood lead group. Among children with high cord-blood lead, concurrent blood lead concentration and sociodemographic characteristics associated with extent of recovery or compensation. 6. IQ and achievement scores at age 10 yr inversely associated with blood lead measured at 24 mo of age.

Study Site: Key References	No.[a]	Blood Lead Assessment	Outcome Assessment[b]	Major Findings
Port Pirie:	723	Maternal (prenatal): $\bar{x}$ ~ 9.3 µg/dL	24 mo: BSID 48 mo: MSCA 7 yr: WISC-R	1. Increased risk of preterm delivery. 2. Reduced head circumference at birth. 3. Indexes of prenatal exposure not related to MDI scores at age 2 yr. 4. Mental Development Index at 24 mo weakly associated with blood lead concentration at 6 mo of age. 5. General Cognitive Index scores at age 48 mo inversely related to integrated average of postnatal blood lead concentrations. 6. IQ at 7 years inversely related to integrated average of postnatal blood lead concentrations.
Baghurst et al., 1987a,b		Cord: 8.3		
Baghurst et al., 1992		6 mo: 14.5 15 mo: 20.9 24 mo: 21.3		
Wigg et al., 1988		36 mo: 19.5 48 mo: 16.4		
McMichael et al., 1988		Integrated postnatal average to age 4 19.1		
Vimpani et al., 1989		Mean lifetime to age 71 9.1		
Cincinnati:	305	Maternal (prenatal): $\bar{x} = 8.0$, SD = 3.7	3 mo: BSID 6 mo: BSID 12 mo: BSID 24 mo: BSID 39 mo: FSLST 48 mo: K-ABC	1. Low birthweight and reduced duration of gestation. 2. Mental Development Index scores at 3 and 6 mo inversely related to prenatal and postnatal blood lead concentrations.
Dietrich et al., 1987a,b; 1989; 1990; 1991; 1992; 1993a,b		Cord: $\bar{x} = 6.3$, SD = 45		

Cincinnati (cont.)	Neonatal (10 day): $\bar{x}$ 4.6, SD = 2.8	60 mo: K-ABC SCAN
Dietrich, 1991; 1992	3 mo: $\bar{x}$ =5.9, SD = 3.4	78 mo: WISC-R
Shukla et al., 1989; 1991	Maximum first yr: $\bar{x}$ =15.9, SD = 8.2	72 mo: BOTMP
	Maximum second yr: $\bar{x}$ = 21.1, SD = 11.4	
	24 mo: $\bar{x}$ = 17.5, SD = 9.2	
	Mean of 3rd yr: 16.3, SD = 7.8	
	Mean of 4th yr: 14.1, SD = 7.3	
	Mean of 5th yr: 11.9, SD = 6.4	

3. Mental Development Index at 12 mo inversely associated (indirectly, via birthweight) with prenatal blood lead concentration.
4. Mental Development Index at 24 mo positively associated with prenatal blood lead concentration.
5. Mental Development Index scores at 3, 6, 12, and 24 mo not associated with postnatal blood lead concentrations.
6. Retarded growth in stature.
7. FSLST scores at 39 mo inversely related to prenatal blood lead concentration.
8. K-ABC scores at age 4 yr inversely related only to neonatal (10-day) blood lead concentration (poorest children only).
9. Poorer central auditory processing abilities associated with higher postnatal blood lead concentrations.
10. K-ABC scores at age 5 yr (simultaneous processing) inversely associated significantly only with mean blood lead concentration in fourth year of life.
11. WISC-R performance IQ and BOTMP scores inversely associated with postnatal lead exposure.

Study Site: Key References	No.[a]	Blood Lead Assessment	Outcome Assessment[b]	Major Findings
Cleveland: Ernhart et al., 1986; 1987; 1989 Ernhart and Morrow-Tlucak, 1987 Morrow-Tlucak and Ernhart, 1987 Ernhart and Greene, 1990 Greene and Ernhart, 1991	359	Maternal (prenatal): $\bar{x} = 6.5$ μg/dL SD = 1.9 Cord: $\bar{x} = 5.8$ 6 mo: $\bar{x} = 10.1$, SD = 3.3 2 yr: $\bar{x} = 16.7$, SD = 6.5 3 yr: $\bar{x} = 16.7$, SD = 5.9 SD = 2.0	Neonatal: NBAS Neonatal: GRBE 6 mo: BSID 6 mo: KID 12 mo: BSID, SICD 24 mo: BSID, SICD 3 yr: SB, SICD 4-10 yr: WPSSI	1. Neurologic soft signs score on GRBE associated with cord-blood lead concentration. 2. Mental Development Index, Psychomotor Development Index, and KID scores at 6 mo inversely related to maternal blood lead concentration during pregnancy. 3. No other associations between either prenatal or postnatal blood lead concentrations and scores or growth indexes.
Sydney: Cooney et al., 1989a,b	318	Maternal (delivery): $\bar{x} = 9.1, 1.3*$ Cord: $\bar{x} = 8.1, 1.4$ 6 mo: $\bar{x} = 15.0, 1.6$ 12 mo: $\bar{x} = 15.4, 1.5$ 18 mo: $\bar{x} = 16.4, 1.5$ 24 mo: $\bar{x} = 15.2, 1.3$ 30 mo: $\bar{x} = 12.8, 1.8$	6 mo: BSID 12 mo: BSID 24 mo: BSID 36 mo: MSCA 48 mo: MSCA	No association between children's scores and any index of lead exposure.

Sydney (cont.)		36 mo: x̄ = 12.0, 1.5 42 mo: x̄ = 10.7, 1.5 48 mo: x̄ = 10.1, 1.4 Multiplicative factors
Kosovo, Yugoslavia:	541	Maternal
Graziano et al., 1990		
Wasserman et al., 1992		

NOTES: [a]Numbers of children recruited into cohort. Numbers included in specific analyses vary with cohort attrition and patterns of missing data.

[b]BSID: Bayley Scales of Infant Development
MSCA: McCarthy Scales of Children's Abilities
NBAS: Neonatal Behavior Assessment Scale
GRBE: Gram-Rosenblith Behavioral Examination
KID: Kent Infant Development Scale
SB: Stanford-Binet Intelligence Scale
WPSSI: Wechsler Preschool and Primary Scales of Intelligence
FSLST: Fluarty Speech and Language Screening Test
K-ABC: Kaufman Assessment Battery for Children
SCAN: Screening Test for Auditory Processing Disorders
SICD: Sequenced Inventory of Communication Development
WISC-R: Wechsler Intelligence Scale for Children-Revised
K-TEA: Kaufman Test of Educational Achievement
BOTMP: Bruininks-Oseretsky Test of Motor Proficiency

TABLE 2-2 Major Cross-Sectional Studies of Low-Dose Exposure

Study	No.	Exposure Index	IQ Measurement	Age at Assessment	Potential Confounders Considered
Needleman et al., 1979	158	Tooth lead	WISC-R	7.4 yr	Mother's age at child's birth, maternal education, father's social class, number of pregnancies, parental IQ
Winneke et al., 1982	52	Tooth lead	German WISC	8.5 yr	Exposure groups matched for age, sex, father's occupational status
Winneke et al., 1983	115	Tooth lead	German WISC	9.4 yr	Age, sex, duration of labor, socio-hereditary background (composite of school type and occupational status of parents)
Smith et al., 1983; Pocock et al., 1987	402	Tooth lead	WISC-R	6 yr	Maternal IQ, quality of marital relationship, family characteristics, parental interest, family size, social class, birthweight, length of hospital stay after birth, sex
Fergusson et al., 1988	724	Tooth lead	WISC-R	8 yr	Maternal education, paternal education, birthweight sex, standard of living, maternal emotional responsiveness, maternal avoidance of punishment, number of weatherboard homes resided in

Study	N	Exposure	Test	Age	Covariates
Hansen et al., 1989a,b	162	Tooth lead	WISC-R	First grade	Number of sibship (birth order), maternal education, maternal age, whether child came home from hospital after mother, jaundice, father's socioeconomic status
Bergomi et al., 1989	237	Tooth lead	WISC-R	7.7 yr	Age, SES, sex
Yule et al., 1981	166	Blood lead	WISC-R	6-12 yr	Age, social class
Lansdown et al., 1986	194	Blood lead	WISC-R	$x = 9.1$ yr	Age, social class
Fulton et al., 1987	501	Blood lead	BAS	6-9 yr	Parents' vocabulary and matrices scores, child's interest score, age father's education, length of gestation, parental involvement in school, class year, days absent from school, height, car and telephone ownership, whether father is unemployed, sex
Hawk et al., 1986	75	Blood lead	SBIT-R	3-7 yr	Maternal IQ, H.O.M.E. (measure of home rearing environment), gender
Silva et al., 1988	579	Blood lead	WISC-R	11 yr	None (no multivariate analysis because blood lead-IQ association not significant in bivariate analyses)

Study	No.	Exposure Index	IQ Measurement	Age at Assessment	Potential Confounders Considered
Hatzakis et al., 1989	509	Blood lead	WISC-R	Primary	Parental IQ, birth order, age, family size, father's age, father's education, occupation, mother's education, alcoholic mother, bilingualism, birthweight, length of hospital stay after birth, walking age, history of CNS disease, history of head trauma, illness affecting sensory function, parental divorce

study. The legends provide additional information to aid the reader in evaluating this issue.

Figure 2-1 displays the IQ scores of children classified by blood lead concentration, which serves as an index of earlier and current lead exposures. Within each cohort, children with lower mean blood lead concentrations scored higher than children with higher mean concentrations, and the decline with increasing exposure was roughly monotonic. The differences among cohorts in overall performance (i.e., height on the ordinate) are substantial, but not surprising, in view of the many differences among studies, including the IQ test used and its appropriateness to the population sampled, the sociodemographic characteristics of the cohort, and the total body lead burden of the children.

Figure 2-2 displays the IQ scores of children in studies that relied on tooth lead as the exposure index. Within each study, children's scores tended to decline with increasing tooth lead. The consistency in the findings is all the more surprising, in view of interstudy differences in the portion of tooth anatomy sampled for analysis (e.g., whole tooth, crown, circumpulpal dentin, primary dentin) as well as in the type and location of teeth obtained for analysis (Smith et al., 1983; Grandjean et al., 1986; Purchase and Fergusson, 1986; Paterson et al., 1988) and, in many cases, the absence of interlaboratory quality-assurance and quality-control procedures.

To a large extent, evaluation of whether a study provides evidence of an association between lead and IQ has traditionally been based on whether the p value related to the association is less than 0.05. A more efficient use of the information from different studies is to assess the consistency in the magnitude of effect sizes. Studies in which the p value was greater than 0.05 can provide evidence that supports the association, if they all report similar effect sizes. The p value associated with an individual study depends on many factors, including sample size, the dispersion of values for exposure and outcome within a cohort, and the precision achieved in measuring exposure, outcome, and covariables. The p value associated with the result of a single study is somewhat less important when the studies are viewed in aggregate.

A meta-analysis of 24 recent comparable studies of the association between lead and IQ (including most of the studies in Figures 2-1 and 2-2) indicated that, under the null hypothesis of no association between lead concentration and IQ, the overall pattern of reported associations is

FIGURE 2-1 (facing page). Blood lead concentrations vs. intelligence test scores. Data from cross-sectional and retrospective cohort studies that relied on blood lead concentration as index of children's exposure. For study to be included, investigators had to present either mean IQ scores for children by blood lead strata or sufficient information about regression of IQ on blood lead to specify regression line (i.e., coefficient for blood lead and intercept of regression line or figure from which they could be determined). Except where noted, scores are adjusted for confounding although control variables vary among studies. Source of information provided for each study depicted is as follows: Yule et al. (1981): mean full-scale WISC-R IQ scores for children in blood lead quartiles. Winneke et al. (1990): WISC scores based on four subscales: Vocabulary, Comprehension, Picture Completion, Block Design; apparently not adjusted for confounding. The group with the highest blood lead levels (mean of 50.1) achieved a mean IQ score of 104.3 (not shown). Fulton et al. (1987): regression of British Ability Scales combined score on blood lead. Based on analysis conducted by Grant and Davis (1989). Lansdown et al. (1986): mean WISC-R IQ scores not presented for complete cohort, only for children stratified by parental occupation (manual vs. nonmanual); apart from stratification on parental occupation, scores are not adjusted for confounding. Hawk et al. (1986): regression of Stanford-Binet IQ scores on blood lead concentration; chunk test evaluating contribution of control and interaction terms was not significant. Schroeder et al. (1985): regression of Stanford-Binet IQ Scores on blood lead concentrations; data represent apparently unadjusted regression of IQ on contemporary blood lead concentration among 6- to 12-year-olds; data selected for inclusion because they are most similar to those from traditional cross-sectional study. Although the model in which blood lead was the only predictor yielded the most precise estimate of the slope of the blood lead-IQ relationship, the slope was reduced from -0.4456 to -0.255 in the model that achieved optimal precision and validity. Source: Adapted from Bellinger and Needleman, 1992.

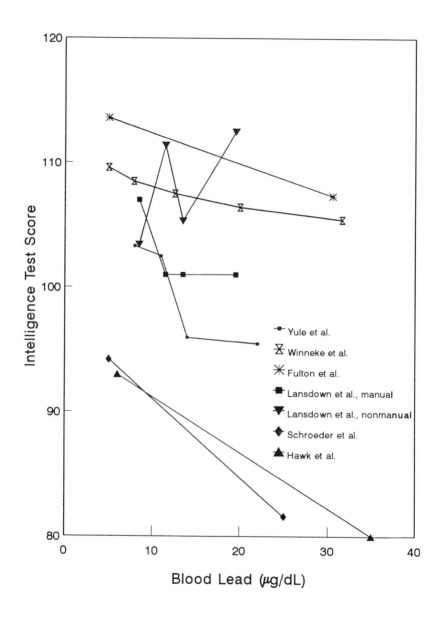

FIGURE 2-1

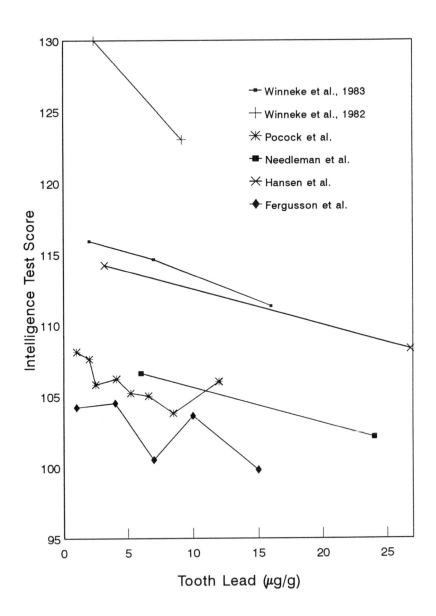

FIGURE 2-2

highly unlikely to have occurred by chance (Needleman and Gatsonis, 1990). Two methods were used to calculate a joint p value: Fisher's method for aggregating p values and Mosteller and Bush's method for calculating a weighted sum of t values. For studies relying on blood lead as the index of children's exposure, both methods yielded a joint p value less than 0.0001. For studies based on tooth lead, the p values were 0.0005 and 0.004. Pooled data from the eight WHO/CEC (World Health Organization and Commission of the European Communities) studies (conducted in Aarhus, Athens, Bucharest, Budapest, Modena, Sofia, Zagreb, and Dusseldorf) resulted in a common regression coefficient of -0.53 ($p < 0.1$, one-tailed) (Ewers et al., 1989; Winneke et al., 1990).

Despite interstudy differences in the ranges of blood lead represented in a cohort, most studies report a 2- to 4-point IQ deficit for each increase of 10-15 µg/dL in blood lead within the range of 5-35 µg/dL. A threshold for that effect of lead is not evident from the reported studies. It is important to note that the effect sizes estimated on the basis of

---

FIGURE 2-2 (facing page). Tooth lead concentrations vs. intelligence-test scores. Data from retrospective cohort studies that relied on concentration of lead in some portion of tooth as index of children's exposure. For study to be included, investigators had to present either mean IQ scores for children by tooth lead strata or sufficient information about regression of IQ on tooth lead to specify regression line (i.e., the coefficient for tooth lead and intercept of regression line or figure from which they could be determined). Except where noted, scores are adjusted for confounding although control variables vary among studies. Source of information provided for each study depicted is as follows: Winneke et al. (1983): verbal IQ scores from German adaptation of WISC; full-scale IQ scores for children in different tooth lead strata were not provided. Winneke et al. (1982): full-scale IQ scores from German adaptation of WISC for matched groups. Pocock et al. (1987): WISC-R scores. Needleman et al. (1979): full-scale WISC-R scores. Hansen et al. (1989): full-scale scores on Danish adaptation of WISC for matched groups. Fergusson et al. (1988): unadjusted full-scale WISC-R scores at age 8 yr; tooth lead values displayed are midpoints of ranges; no range is provided for highest tooth lead strata (12+), so 15 µg/g was chosen; identical pattern was evident in WISC-R scores at 9 yr of age (not shown); adjusted scores were not provided. Source: Adapted from Bellinger and Needleman, 1992.

prospective longitudinal studies and cross-sectional studies are essentially identical. In a recent meta-analysis of studies reporting on cognitive function at school-age, Schwartz (1992a, in press (a)) calculated the IQ decline over the blood lead range of 10 to 20 µg/dL to be 2.32 points (standard error of 1.27) for longitudinal studies and 2.69 points (standard error of 1.28) for cross-sectional studies. The public-health significance of such an effect size has stimulated spirited debate.

The public health significance of an effect size of this magnitude has stimulated spirited debate. Three issues warrant consideration. First, SEM is a concept that pertains to the performance of an individual, not a group. Specifically, it defines the range, centered around a subject's observed score, within which his or her true score is likely to lie. Thus, SEM is not germane in interpreting the importance of group differences in mean score. Second, a property of statistical distributions is that a small difference in mean score between two groups results in substantial differences in frequency of extreme values between the two distributions. The distributional implications of small changes in population mean score have been confirmed by analyses of several lead-study data sets (Needleman et al., 1982; Bellinger et al., 1989a; Davis, 1990). Third, small differences in IQ have been associated with differences in measures of socioeconomic success, such as wages and educational attainment (Griliches, 1977).

Although the search for markers of increased vulnerability has been carried out only with post hoc analyses, results of several studies suggest that children in lower social strata (or whose parents have manual occupations) express an IQ deficit at lower exposures than do children in higher social strata (Harvey et al., 1984; Winneke and Kraemer, 1984; Rabinowitz et al., 1991). Sociodemographic differences are thought to account for the discrepancy between the results of the first (Yule et al., 1981) and second (Lansdown et al., 1986) London studies. Not all studies report socioeconomic differences in vulnerability, however. In one study, the association between increased tooth lead and lower IQ was more prominent among boys than girls (Pocock et al., 1987). The findings should be viewed as preliminary, but they are consistent with patterns reported in two of the prospective studies (Dietrich et al., 1987a; Bellinger et al., 1990). In two recent studies, however, the inverse association between lead level and cognitive performance was stronger among girls than boys (Rabinowitz et al., 1991; Leviton et al., 1993), leaving the issue of sex differences in vulnerability unresolved.

Several obstacles impede efforts to discern whether specific neuropsychologic deficits are associated with higher lead exposures. First, differences among investigators in interests and experience, as well as national differences in assessment strategies and approaches, have contributed to interstudy differences in the instruments used and the ages at which children were assessed. Second, within an individual study, the instruments used to assess function in different cognitive domains vary in reliability and sensitivity. If children with different exposures perform differently on test A but not on test B, it might be difficult to determine whether the contrast is attributable to lead-associated effects on the skills underlying test A but not those underlying test B or to the superiority of test A in psychometric properties. Third, the specific manifestations of lead's cognitive toxicity might vary with characteristics of a cohort, such as socioeconomic status or other markers of the types of developmental support available to children. For instance, children in lower social strata often begin to manifest language deficits in the second year of life that are attributed to a relative lack of environmental support for the types of linguistic skills assessed in standardized tests. The increased vulnerability of verbal function in such children might make this aspect of cognition most sensitive to toxic exposure. In children from higher social strata, where greater emphasis might be placed on the development of primarily verbal academic skills, this aspect of cognitive function could be more protected. Toxicity might be expressed in other ways, such as visual-spatial or visual-motor integration. Fourth, differences across studies in the cognitive domains found to be associated with lead might reflect differences in the exposure histories of the children in various cohorts and to differences in the exposure index used. Some cognitive functions might be more strongly associated with exposure within the first 2 years of life, and others with later exposure (Shaheen, 1984). For still other functions, the important contrast could be between cumulative and acute exposure (e.g., Winneke et al., 1987, 1988).

As noted in the discussion of the impact of lead on IQ, the importance of p values should not be magnified in assessing the consistency across studies in the association of specific cognitive functions with lead. Numerous studies showing similar effect sizes, some of which might not be statistically significant, are more persuasive than a set of studies showing discrepant effect sizes with similar p values.

There is relatively little consistency across studies in terms of whether verbal IQ or performance IQ is more strongly associated with lead exposure. Some studies report stronger associations for verbal IQ or surrogate scores (Needleman et al., 1979; Ernhart et al., 1981; Yule et al., 1981; Bergomi et al., 1989; Hansen et al., 1989a,b), and others for performance IQ (Marecek et al., 1983; Shapiro and Marecek, 1984). In some studies, size of exposure was significantly associated with both scales (Hatzakis et al., 1989), and in others, with neither scale (Smith et al., 1983; Winneke et al., 1983; Lansdown et al., 1986; Fergusson et al., 1988; Silva et al., 1988). Similar inconsistency has been reported in the results of IQ testing conducted at school-age in the prospective studies (Baghurst et al., 1992; Bellinger et al., 1992; Dietrich et al., 1993a). Several studies have noted significantly lower reading scores (primarily word-reading) among children with larger exposures (Yule et al., 1981; Fulton et al., 1987; Fergusson et al., 1988; Needleman et al., 1990); other studies have noted similar, but nonsignificant, trends (Ernhart et al., 1981; Smith et al., 1983; Silva et al., 1988). Spelling deficits have also been reported (Yule et al., 1981; Fergusson et al., 1988). Some studies report significant associations between lead exposure and mathematical skills (Fulton et al., 1987; Fergusson et al., 1988); others do not (Yule et al., 1981; Smith et al., 1983; Lansdown et al., 1986).

In several studies, children with larger lead exposure did poorly on assessments of visual-spatial or visual-motor skills, with deficits apparent on figure reproduction, visual retention, mazes, eye-hand coordination, and construction tasks (Winneke et al., 1988; McBride et al., 1982; Bellinger et al., 1991a; Hansen et al., 1989a,b). Analysis of the pooled data from the WHO/CEC studies (Ewers et al., 1989) indicated a significant positive association between errors on the Bender-Gestalt Test (German scoring system) and blood lead concentration, particularly on the more difficult trials when perceptual distractions were introduced.

A more consistent finding across studies is an inverse relationship between children's lead concentration and the adequacy of their performance on simple and especially choice reaction-time tasks (Needleman et al., 1979, 1990; Winneke et al., 1983, 1989, 1990; Hunter et al., 1985; Hatzakis et al., 1989; Raab et al., 1990). In the WHO/CEC studies, blood lead concentration was positively associated with errors and negatively associated with hits on a serial-choice reaction-time task (Ewers et al., 1989). Lead exposure was not significantly associated with performance on a delayed-reaction-time task in this set of studies. Larger exposure has also been linked to poorer performance on tests such as the Toulouse-Pieron cancellation test (Bergomi et al., 1989), the

Trail-Making Test, Stroop Test, the Talland Letter Cancellation Test, and the Wisconsin Card Sorting Test (Bellinger et al., in press). If the assumption that such tasks assess children's attention skills is correct, these data are consistent with other findings, based on teachers' ratings, that children with larger exposures are less attentive in the classroom (Needleman et al., 1979; Yule et al., 1984; Hatzakis et al., 1987; Silva et al., 1988; Thomson et al., 1989). Blood lead concentration was not significantly associated with teachers' ratings of classroom behavior in the WHO/CEC studies, however (Ewers et al., 1989). Except for the finding of Hansen et al. (1989a,b) of greater off-task behavior during the continuous-performance task, direct observations of children have not demonstrated behavioral differences between groups of children with varied magnitudes of lead exposure (Bellinger et al., 1984b; Harvey et al., 1984).

Byers and Lord (1943) reported that 19 of 20 children with asymptomatic lead poisoning failed to achieve adequate progress in school, despite IQs in the normal range. Their difficulties were attributed to behavioral dysfunctions, such as distractibility and impulsivity. Although those observations are generally credited with originating studies on so-called subclinical effects of lead exposure, relatively few of the more recent studies have examined performance in school as an outcome variable, apart from collecting teachers' ratings of children's classroom behavior.

The limited data available are generally consistent with the hypothesis that children with greater lead burdens not only perform worse on laboratory and psychometric tests of cognitive function, but also are more frequently classified as learning-disabled and make slower progress through the grades. For instance, in a followup study of a subset of 141 children in the cohort originally identified by Needleman et al. (1979), dentin lead concentrations greater than 20 parts per million (ppm) were associated with increased rates of referral for remedial academic help and with grade retention during the late elementary-school years (Bellinger et al., 1984b). In a cross-sectional study of 200 second-grade Scandinavian children, the risk (adjusted odds ratio) for learning disability among children with circumpulpal-dentin lead concentrations greater than 16 ppm was 4.3 (the reference was the rate among children with concentrations less than 5 ppm) (Lyngbye et al., 1990). Including children with a variety of medical risk factors reduced the odds ratio, but the risk of learning disability among children with high dentin lead remained double the reference risk.

An assessment of the prevalence of lead-associated learning disabilities over a much longer followup interval was reported by Needleman et al. (1990). At age 18-19 years, children with high dentin lead concentrations had significantly higher rates of reading disability (at least two grades below expected) and failure to graduate from high school; the adjusted odds ratios were 5.8 and 7.4, respectively, when the prevalence among children with dentin lead concentrations below 10 ppm was used as the reference.

In children, early neurobehavioral and other developmental effects have been reported at blood lead concentrations of 10 μg/dL or even lower (and equivalent concentrations in other tissues). Figure 2-3 shows a nonparametric smoothed curve of full-scale IQ versus dentin lead concentration with covariates controlled for. The figure comes from a reanalysis of the data of Needleman et al. (1979) by Schwartz (in press). The analyses of Needleman and co-workers have recently been criticized. It has been suggested that their finding of a significant association between full-scale IQ and dentin lead followed from three critical choices: their exclusion of subjects with characteristics that they felt might be strongly related to the outcome (such as hospitalization for head injuries or residence in non-English-speaking homes), their use of external age adjustment rather than direct control for age in the regression, and their method for assigning subjects to lead-exposure groups. The reanalysis addressed recent criticisms of the original analysis by

- Including all the subjects, instead of using the exclusion criteria of Needleman and co-workers.
- Controlling directly for age in the regression model, instead of using indirect standardization.
- Using the mean dentin lead in each child as the exposure index, instead of a set of categorization rules that discarded discordant values.

The reanalysis also controlled for additional covariates. Dentin lead concentrations were found to be more highly significantly related to full-scale IQ than in the original analysis. Figure 2-3 indicates that the covariate-adjusted association continued to the lowest dentin lead concentration found in the sample, 1 ppm. Although that cannot speak to

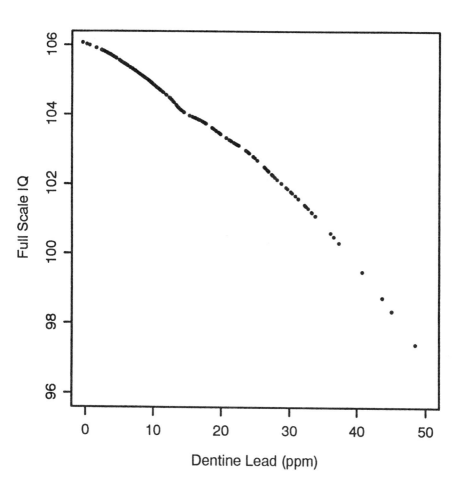

FIGURE 2-3 Nonparametric smoothed plot of full-scale IQ vs. dentin lead from Needleman et al. (1979). After controlling by regression for age, maternal IQ, maternal age, mother's and father's education, mother's and father's SES, number of siblings and hospitalization for lead poisoning. None of subjects excluded in original analysis were excluded from reanalysis. Source: Schwartz, in press. Reprinted with permission from *Neuro-Toxicology*; copyright 1992, Intox Press.

effects at lower concentrations, the rarity of concentrations below 1 ppm in industrial societies suggests a lack of an effective threshold. Schwartz (in press) presents a plot that indicates that smoothing did not distort the relationship. Schwartz also reports a reanalysis of the data of Bellinger et al. (1991a). The Schwartz reanalysis—in addition to addressing the question of the impact of the exclusions, age-control method, and definition of exposure in the original paper of Needleman et al. (1979)—also went to some lengths to examine the sensitivity of the conclusions to those and other factors. The regression coefficients and standard errors for the baseline model (which controlled for age, used mean dentin lead as the exposure index, and used no exclusionary rules) were compared with those for a number of different models. Some additional covariates were included, and some of the original Needleman exclusionary rules were used.

The association between dentin lead and full-scale IQ was insensitive to those changes. To ensure that the association was not driven by a few influential observations, Schwartz used M estimation, a technique that assigns lower weight to points that are far from the predicted regression line to reduce the possible influence of a few anomalous points. Bootstrapping, which calculates mean regression coefficients and confidence intervals by repeatedly resampling observations from the original data set, was also used; it yields inferences that are less sensitive to any assumptions about the distributional properties of the variables and parameters. Both techniques gave results essentially identical with those of the baseline model. Robust variance estimates also yielded the same results. The nonparametric smooth curve mentioned above (Figure 2-3) also indicates that the relationship is not driven by a few selected observations.

Schwartz also examined the association between dentin lead and other outcomes. Those outcomes might be expected to covary with IQ, and an association with them further indicates that the IQ results are not anomalous in these data. In models that control for all the covariates used in the IQ regressions, dentin lead was associated with Teacher Rating Score, with Piaget Mathematics and Reading Scores, with Seshore Rhythm, and with other neurodevelopmental outcomes. The findings suggest that the association between dentin lead and intellectual development in the data of Needleman and co-workers is strong and robust. Bellinger et al. had reported that 20-month blood lead concentrations were associated with 57-month McCarthy General Cognitive Index in their prospective study of lead exposure. Using least-square means, they showed that children with blood lead of 3-9 µg/dL had significantly lower McCarthy scores than children with blood lead below 3 µg/dL. Figure 2-4 shows a covariate-adjusted nonparametric smoothing of the McCarthy Scores for those children versus blood lead concentrations from Schwartz's reanalysis. A continuous dose-dependent decline is seen to start at 1 µg/dL. Figure 2-5 shows the covariate-adjusted Bayley MDI scores at age 18 months for three categories of cord lead concentration, as reported from the Boston study. More recently, Schwartz (in press), using hockeystick regression, has demonstrated a threshold estimate below 1 µg/dL in the relationship between McCarthy Global Cognitive Index and blood lead in these data. Those studies support the general conclusion that there is growing evidence that there is no effective threshold for some of the adverse effects of lead.

Another neurobehavioral end point is evident in Figure 2-6, which shows the percent of children with hearing levels worse than the refer-

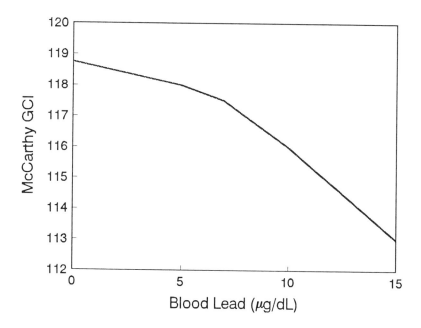

FIGURE 2-4 Nonparametric smoothed plot of McCarthy Global Cognitive Index vs. blood lead at 24 months of age in Boston prospective lead study (Bellinger et al., 1991a). Source: Schwartz, in press. Reprinted with permission from *NeuroToxicology*; copyright 1992, Intox Press.

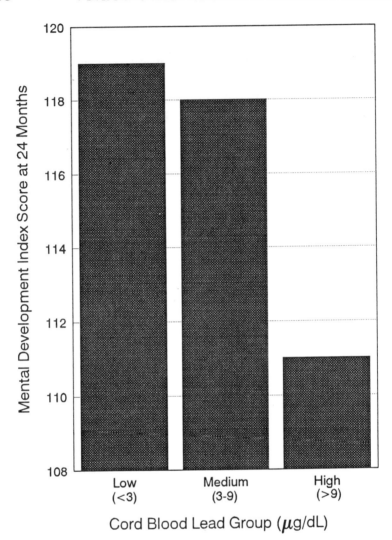

FIGURE 2-5 Mean Bayley Mental Development Index in children aged 24 months, by umbilical cord lead group, after adjustment for covariates, from study of Bellinger and co-workers (1987a,b). Source: Adapted from Schwartz, in press. Reprinted with permission from *NeuroToxicology*; copyright 1992, Intox Press.

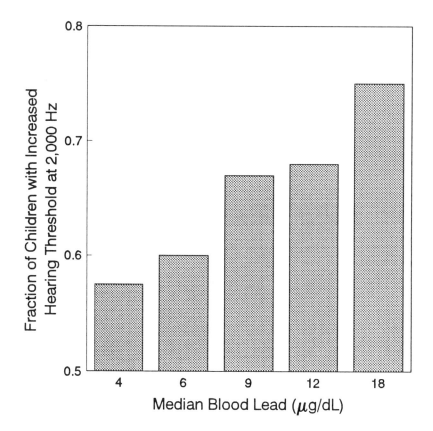

FIGURE 2-6 Fraction of children with hearing worse than reference level of quintiles of blood lead concentration, after adjustment for covariates. Data from Schwartz and Otto (1991). Source: Adapted from Schwartz, in press. Reprinted with permission from *NeuroToxicology*; copyright 1992, Intox Press.

ence level, by quintiles of blood lead concentration, after adjustment for covariates. The data are from the Hispanic Health and Nutrition Examination Survey, as reported by Schwartz and Otto (1991). The effects clearly continue to well below 10 µg/dL.

Considerable interest has focused on the persistence of the cognitive deficits seen in lead. The longest followup study, published recently by Needleman and co-workers (1990), showed that some deficits persisted and showed a dose-dependent relationship with lead exposure. Figure

2-7 shows the fraction of children with reading disability, by quartile of dentin lead concentration, after adjustment for covariates.

In the paper of Schwartz and co-workers (1986), children's stature was associated with blood lead concentrations. A hockeystick regression analysis found no evidence of a threshold down to the lowest blood lead concentration in the data (2 μg/dL). At lower ages, Shukla and colleagues (1987, 1989) found an association between integrated postnatal blood lead and child's stature at 33 months. Figure 2-8 shows that relationship, after adjustment for covariates.

Neurotoxic effects of lead in addition to effects on cognition and other neurobehavioral measures in children have been documented in both the central nervous system and the peripheral nervous system (PNS) of both adults and children.

In both lead workers and lead-exposed children, one noninvasive,

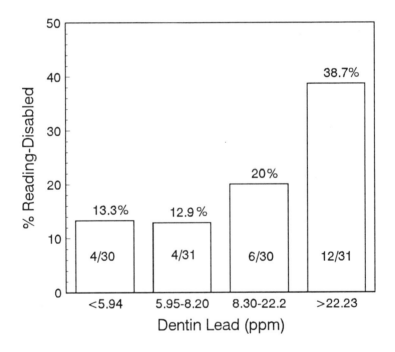

FIGURE 2-7 Percentage of children with reading disability (defined as two or more grades below expected) by quartiles of dentin lead concentration after adjustment for covariates. Data from long-term followup of Needleman and co-workers (1990). Source: Schwartz, in press. Reprinted with permission from *NeuroToxicology*; copyright 1992, Intox Press.

useful measure of PNS injury is the reduction of conduction velocity in some sensory and motor nerves. By and large, lead workers show impairment of nerve conduction velocity at relatively higher concentrations of blood lead than those associated with either childhood lead neurotoxicity or that related to other toxic end points in adults. Nerve conduction-velocity impairment appears not to be a particularly sensitive measure of neurotoxicity in adults, as it is a measure that reflects advanced manifestation of demyelinating injury involving Schwann cells.

Effects of lead on peripheral nerve function in children are also known, although not as well studied as in adults. Studies of inner-city children (Feldman et al., 1973a,b; 1977) and children residing in smelter communities (Landrigan et al., 1976; Englert, 1978; Winneke et al., 1984; Schwartz et al., 1988) have been reported. Multiple statistical analyses (Schwartz et al., 1988) of nerve conduction-velocity data obtained from a group of asymptomatic smelter-community children described earlier (Landrigan et al., 1976) demonstrated a threshold in children for nerve conduction-velocity reduction of blood lead concentration ranging from 20 to 30 μg/dL, depending on the statistical analysis. One complication with nerve conduction velocity as a toxicity measure is that a dose-dependent biphasic response can be identified, i.e., a U-shaped dose-effect curve across studies and across a broad range of blood lead concentration (e.g., Schwartz et al., 1988; Winneke et al., 1989).

Various assessments of neurophysiologic end points have involved various evoked-potential testing, particularly by Otto (Benignus et al., 1981; Robinson et al., 1985; Otto, 1989). The salient points of the various studies are as follows:

• Various evoked-potential tests are measures of CNS perturbations in young children, even though some inconsistencies across time and stage of neurologic development suggest that multiple mechanisms are involved.
• Linear dose-effect data were reported in connection with conditioned slow-wave voltage changes in children, brain-stem evoked-potential latencies, pattern-reversal evoked-potential (PREP) latencies, and PREP amplitude.
• Such measures seem to be minimally affected by social and cultural factors that complicate psychometric studies.
• Although much of the evoked-potential information examined by Otto and others might not have clear clinical connections, except for that

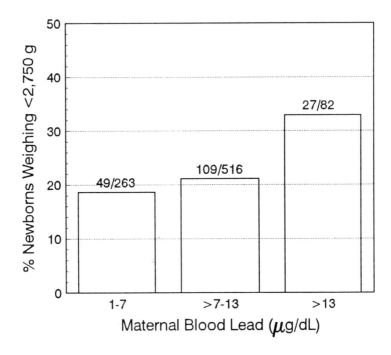

FIGURE 2-8 Percentage of newborns weighing less than 2,750 g by maternal blood lead category, after adjustment for covariates. Data from Dietrich (1991). Source: Schwartz, in press. Reprinted with permission from *NeuroToxicology*; copyright 1992, Intox Press.

linked to hearing impairment (Schwartz and Otto, 1987), any CNS perturbations that occur in developing children should be regarded with the utmost concern.

CARDIOVASCULAR EFFECTS

Hypertension and Pregnancy

Ever since Schedoff and Porockjakoff drew attention to the association of high blood pressure and eclampsia in 1884, there has been increasing interest in this relationship. In normal pregnancy, despite the 30-40% increase in blood volume and cardiac output, arterial pressure falls, because of a decrease in peripheral vascular resistance. A fall early in gestation lasts until around weeks 18-24 of pregnancy. Between weeks 24 and 26, blood pressure increases to a plateau; it then stays steady until delivery. Women who do not follow that usual pattern and are hypertensive during pregnancy have a higher risk of adverse pregnancy outcome. The reported prevalence of hypertension ranges from 3.5% to 23.6% (Underwood et al., 1967; Russell et al., 1968; Naidoo and Moodley, 1980; Huisman and Aarnoudse, 1986); the lowest of those figures is for a population mostly of Caucasians, and the highest is for a group of nulliparous women at risk of poor pregnancy outcome. The estimates of prevalence have suffered from lack of uniform definition and of standardized measurements of blood pressure.

Hypertension is the disease most often associated with fetal growth retardation (IOM, 1985). Low and Galbraith (1974) attributed 27% of intrauterine growth retardation with an identifiable cause to severe preeclampsia, chronic hypertensive vascular disease, or chronic renal disease. Breart et al. (1982) found that intrauterine growth retardation occurred in only 3% of births when the diastolic blood pressure was less than 90 mm Hg, 6% of births at 90-110 mm Hg, and 16% of births when it was higher than 110 mm Hg.

Lead readily crosses the placental barrier during the entire gestational period, and its uptake appears to be cumulative until birth (ATSDR, 1988). McMichael et al. (1986) showed that preterm deliveries and reduction in birthweight were significantly related to maternal blood lead at delivery. Those findings have been documented in other studies in the United States and other countries. The prenatal effects are minimal or disappear and thus do not show compromised neurologic functioning in children.

The consistency of the types of adverse pregnancy outcomes that have been related to hypertension and to blood lead in separate studies is striking, but needs to be better integrated. Rabinowitz et al. (1987) are the only investigators who have reported on the relationship of pregnancy with blood pressure and blood lead during pregnancy. They studied 3,200 live births in Boston by white, middle-class women. They examined umbilical-cord blood lead and reported a significant association with blood pressure at delivery and the presence of hypertension during pregnancy. Schwartz (1991) has reported an association between blood lead and blood pressure in females in a nationally representative sample.

A comprehensive review of both human and animal studies has been published (Boscolo and Carmignani, 1988).

Animal Models of Lead and Blood Pressure

In the last decade, a substantial body of animal data have associated in vivo lead exposure with increased blood pressure in animals. Increases in blood pressure in the rat have been documented by Victery et al. (1982), Iannacone et al. (1981), Perry and Erlanger (1978), Carmignani et al. (1983), Webb et al. (1981), Evis et al. (1987), Boscolo and Carmignani (1988), Bodgen et al. (1991), Nakhoul et al. (1992), and Lal and co-workers (1991). In many of these studies, the blood lead levels involved were quite low. Figure 2-9, for example, shows a plot of the

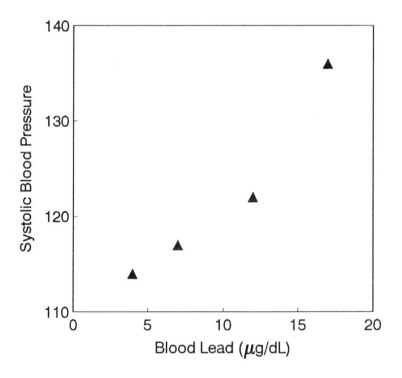

FIGURE 2-9 Response of blood pressure to blood lead concentrations in rats (Boscolo and Carmignani, 1988). Source: Schwartz, 1992b. Reprinted with permission; copyright 1992, CRC Press.

data from Boscolo and Carmignani (1988), taken from the review of Schwartz (1992b). Other studies have documented similar increases in pigeons (Revis et al., 1981).

A common finding in these studies was increased reactivity to alpha-adrenergic stimulation, prolonged response to norepinephrine stimulation, and reduced effectiveness of isoproterenol in lowering blood pressure. These all point to a mechanism involving the modulation of the calcium messenger system that regulates blood pressure. This system is disturbed by lead in other organs as well.

In vitro studies have also supported the role of lead in increasing blood pressure. Isolated tail arteries have shown increased contractile response in studies by Piccini et al. (1977), Chai and Webb (1988), Skoczynska et al. (1986), Carmignani et al. (1983), Iannocone et al. (1981), and Webb et al. (1981). These studies also document increased responsiveness to alpha-adrenergic stimulation. Chai and Webb (1988) report that the contractile response to lead in an isolated rabbit tail artery was increased by a protein kinase C stimulant, and decreased in the presence of a protein kinase C inhibitor. This again suggests the centrality of the calcium messenger system. This system regulates tone in the vascular periphery of humans as well as animals.

Population-Based Epidemiology Studies

Recent epidemiology studies have generally supported the conclusion that the animal results are generalizable to humans. Positive associations between blood lead levels and blood pressure have been reported in essentially all studies, and most of them have reported significant results. The largest studies were the British Regional Heart Study and the NHANES II study. The similarity in the regression coefficients between those two studies has been noted by EPA in the air lead criteria document (EPA, 1986a) as well as by Pocock et al. (1988). Figure 2-10 shows the estimated changes in systolic blood pressure for a change in blood lead from 10 μg/dL to 5 μg/dL from 11 recent studies of the association between blood lead and blood pressure (Schwartz, 1988; Orssaud et al., 1985; Kroumhout, 1988; Pocock et al. 1988; Elwood et al., 1988a,b; Neri et al., 1988; Moreau et al., 1982; de Kort et al., 1987; Sharp et al., 1988). As can be seen, positive and moderate consistent effects are seen in the studies. Other studies have also report-

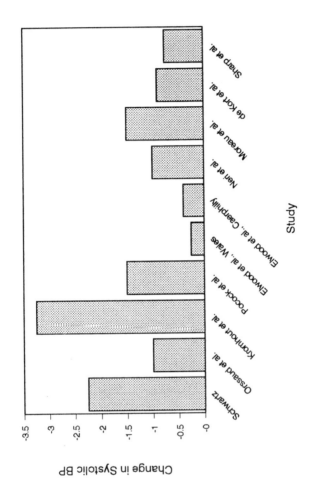

FIGURE 2-10 Reported changes in systolic blood pressure associated with a decrease in blood lead from 10 to 5 µg/dL. Bars: (1) Schwartz, 1988; (2) Orssaud et al., 1985; (3) Kromhout et al., 1985, Kromhout, 1988; (4) Pocock et al., 1985, 1988; (5) Elwood et al., 1988a,b, Welsh Heart Program; (6) Elwood et al., 1988a,b, Caerphilly Collaborative Heart Disease Studies; (7) Neri et al., 1988; (8) Moreau et al., 1982; (9) de Kort et al., 1987; de Kort and Zwinnis, 1988; (10) Sharp et al., 1988. Source: Adapted from Schwartz, 1992b.

ed significant associations, including Weiss et al., (1986), Moller and Kristensen (1992), Egeland et al. (1992), Apostoli et al. (1992), and Hu (1991). No association was reported by Grandjean et al. (1989), and a mixed result, with positive associations in one model, and negative associations in a model with multiple divalent cations, was reported by Staessen et al. (1991). Overall, a considerable majority reported significant associations. Combined with the strong animal model, mechanistic results, and the moderate concordance of effect size, this suggests overwhelming evidence for the causality of the association. This conclusion was also the consensus of the International Conference on lead in blood pressure (Victery et al., 1988), and of EPA's external Science Advisory Board.

Given the estimated changes in blood pressure from Figure 2-10, a study would need extremely large sample sizes to test whether the expected consequences of increased blood pressure on myocardial infarctions occur. No study done to date has had the power to detect relative risks of 1.05. Two studies have focused on intermediate, and more common, cardiovascular end points. Kirkby and Gyntelberg (1985) reported that lead exposure was associated with electrocardiogram changes associated with ischemic heart disease. This was confirmed in a general population study by Schwartz (1991).

MECHANISMS OF TOXICITY

The search for mechanisms of lead's toxic actions in human and experimental animal populations must depend to some extent on the level of the mechanistic explanations being sought. For example, histopathologic, physiologic, cellular, and subcellular and molecular levels of mechanistic explanation have been invoked in numerous papers on human and experimental lead toxicity. Eventually, all the mechanistic explanations are traced to lead's functions at the molecular level in various cell types and in various subcellular components. Investigators have long sought a global, unifying molecular mechanism of lead's toxic action at all sites in the human body. The diversity of toxic effects has made such an explanation extremely difficult to find, and it is the molecular-level mechanism that is of principal interest in this report.

The consequences of lead action on organ and tissue function are also highly correlated with reproducible lead-induced dysfunction in cell culture models of neurons and glia, brain capillary epithelium, bone, and several other pertinent cell types (Tiffany-Castiglioni, 1993; Goldstein, in press; Pounds et al., 1991). Finally, the many toxicological effects of lead are well supported in hypothesis and theory by lead-dependent perturbation of critical physiological, biochemical, and molecular events including signal transduction processes, gene expression, mitochondrial function, etc. (Goering, in press; Regan, in press; Shelton, 1993; Simons, in press).

Most attempts to elucidate or define the mechanism of neurotoxic action of lead on cognition and related measures of brain function include altered development and maintenance of the neural network by lead as a vague, but central thesis. The concept that lead might provoke local or global changes in brain development, architecture, organization, and function is reasonable and supported directly and indirectly by the studies from several laboratories. Two broad classifications of mechanisms have been proposed by Silbergeld (1992). First, are the neurodevelopmental mechanisms that result in persistent and irreversible changes in the architecture of the nervous system. The best example of this developmental mechanism is the neural cell adhesion molecule discussed below. Second, lead interferes with signal transduction processes, especially those associated with neurotransmitter function, which may be reversible. Although these two broad mechanisms may overlap if the neuropharmacologic effects of lead contribute to the developmental mechanism, they provide a useful framework to organize information.

Definition of mechanism of action: To assess appropriately the literature developing on mechanisms of action for lead toxicity and neurotoxicity, it is important to note that the cellular and molecular effects of lead are parallel to the effect of lead on nervous system function in humans. That is, at any level of biologic organization, lead toxicity manifests a broad continuum of toxicity from overt at higher levels to multifactorial recondite toxicities at lower exposure levels. A similar broad continuum of toxicities is observed at the cellular level. Thus, it should not be expected that the actions of lead on a single cellular or molecular process will provide an adequate description of the mechanism of action.

In addition, there is not agreement among investigators as to what constitutes a mechanism of action as the definition of mechanism of action, like the definition of beauty, lies in the eye of the beholder. For example, the underlying process responsible for poor school performance may be best related to another behavioral outcome, such as visual-moter integration. At another, but more remote level, perturbation of signal transduction or gene expression may be responsible for changes in neuronal development and the hard-wiring of the nervous system that may underlie the changes in behavior. At yet another level, interaction of lead with critical sites on specific proteins may explain the effects of lead on signal transduction processes.

Neurotoxicity: The principal neuropathologic feature of acute lead encephalopathy is interstitial edema. Several lines of investigation implicate functional changes in the permeability and barrier properties of the capillary endothelium (Bressler and Goldstein, 1991). These changes in endothelium function may be mediated by the effects of lead on astrocytes possibly through altering calcium homeostasis or by activation of protein kinase C (Gebhart and Goldstein, 1988).

Neurodevelopmental toxicity: The neural cell adhesion molecule (NCAM) is a complex of three polypeptides which regulates many neurodevelopmental processes including neuronal fiber outgrowth and synapse formation (Edelman, 1986). The extracellular domain of the NCAM complex is modified by the addition of sialic acid moieties, with the embryonic form more polysalicylated than the adult form. The sialic acid content determines the strength of interactions between NCAMs on adjacent cells. Chronic lead exposure decreases the rate of NCAM desialylation and the conversion from the embryonic to adult stage in the rat cerebellum (Cookman et al. 1987; Regan, 1989, in press). It is logical that the impaired NCAM desialylation may induce a dys-synchrony in normal cerebellar development, with subsequent altered neuronal structuring contributing reduced fine motor skills and other manifestations of toxicity. However, the biochemical and cellular mechanisms by which lead impairs desialylation remain to be clarified.

Many studies have evaluated the effects of lead on selected parameters of neurotransmitter function and the electrophysiological consequences. Changes in neurotransmitter levels, turnover, and release are well documented in numerous experimental systems, including the neuromuscular junction, synaptosomal preparations, brain tissue slices and cultured neurons. Although there are inconsistencies and contradictions in the findings among these studies, the conclusions are remarkably consistent when the differences in experimental design and the experimental system are considered (Bressler and Goldstein, 1991; Minnema, 1989; Silbergeld, 1992). Chronic exposure to low levels of lead enhances the basal or spontaneous release of various neurotransmitters from almost all systems investigated. For example, lead concentrations of 40 μg/dL increased the frequency of miniature end plate potentials, but did not affect the presynaptic nor the end plate potential after direct stimulation (Atchison and Narahashi, 1984; Cooper et al., 1984; Manalis and Cooper, 1973). In contrast, lead at higher concentrations blocked the evoked release of neurotransmitters in both the peripheral and central nervous system preparations.

Interactions of lead with the calcium messenger system have received considerable attention during the last twenty years. This attention is the result of the physico-chemical similarity between Pb^{2+} and Ca^{2+} and the ubiquitous role of calcium ions as intracellular messengers for transducing electrical and hormonal signals. The interaction of lead with Ca^{2+} homeostasis and the calcium messenger system has been reviewed in detail (Pounds, 1984; Pounds et al., 1991; Simons, in press; Bressler and Goldstein, 1991).

The concentration of free cytoplasmic calcium ion, $[Ca^{2+}]i$, is normally maintained between 50 and 150 nM by the calcium homeostasis system. An appropriate hormonal or electrical signal at the plasma membrane is transduced to a cytoplasmic Ca^{2+} signal by increasing the $[Ca^{2+}]i$ in one or more parts of the cell. Lead interferes with the generation of a Ca^{2+} signal in many cells and nerve terminals. Recent work has extended this understanding by demonstrating that Pb^{2+} inhibited Ca^{2+} entry when calcium channels were opened by depolarization (Simons and Pocock, 1987).

Cytoplasmic Ca^{2+} signals are received by a variety of Ca^{2+} receptor proteins including calmodulin, protein kinase C, calcimedins, parvalbumins, troponin C and many others. Some of these Ca^{2+} receptor proteins are specific to certain cell types, while others are ubiquitous. Two of the most versatile and ubiquitous Ca^{2+} receptor proteins are calmodulin and the protein kinase C family. The calmodulin-mediated responses are typically of brief duration. Typical calmodulin-mediated functions include neurotransmitter release, endocrine and exocrine secretion, etc. Protein kinase C is activated by Ca^{2+} and a lipid metabolite produced by phosphoinositol metabolism, diacylglycerol. Protein kinase C activates protein kinase and phosphatases with both a broad and narrow spectrum of protein substrates. Protein kinase C-mediated responses are typically of longer duration than calmodulin-mediated responses and include cell division and proliferation, cell-cell communication, organization of the cytoskeleton, etc. Lead can perturb the function of these Ca^{2+} receptor proteins directly by substituting for Ca^{2+} with more or less activity, or indirectly by interfering with the generation or removal of the Ca^{2+} signal. For example Pb^{2+} will effectively and functionally displace or substitute for Ca^{2+} in calmodulin and other receptor proteins tested to date (Habermann et al., 1983; Fullmer et al., 1985; Richardt et al., 1986). High levels of calmodulin are particularly associated with the nerve terminals where calmodulin-dependent phosphorylation regulates neurotransmitter release. The inappropriate, or prolonged activation of calmodulin by Pb^{2+} rather than by Ca^{2+} would logically explain the increased spontaneous neurotransmitter released observed by many investigators.

Protein kinase C (PKC) is not a single protein, but a family of isozymes, most of which are calcium activated. Protein kinase C has a profound effect on cell function, especially the regulation of cell growth and differentiation. Markovac and Goldstein (1988a,b) demonstrated that very low levels of lead substituted for calcium in the activation of protein kinase C enzyme activity. Unfortunately, there is not a clear understanding as to the mechanism by which Ca^{2+} activates protein kinase C. Thus the exact biochemical mechanism by which lead activates PKC is only speculation. Nevertheless, the activation of PKC by lead has been confirmed in several laboratories using other tissue or cellular preparations, and thus different PKC isozyme patterns (Goldstein, 1993). Very high levels of lead which are not reasonably expected in vivo are required to inhibit PKC activity. Current evidence correlates activation of PKC activity with functional changes in brain microvascular formation in culture after activation by lead. Similar persistent changes in neuronal activity could underlie the more subtle effects of lead on neuronal function. Thus, the Pb^{2+}-protein interactions with Ca^{2+} receptor proteins and other proteins, such as those of heme biosynthesis, are beginning to be understood (Goering 1993).

Lead has diverse and complex actions on the calcium messenger system, emphasizing the importance of this pathway as a key molecular and cellular target of lead toxicity. Although the effects of lead on these cellular and molecular processes is clearly established, the causal link between these effects and the subtle effects of chronic, low-level lead exposure is difficult to define with experimental rigor.

Effects on Heme Biosynthesis and Erythropoiesis

EPA (1986a) extensively reviewed the effects of lead on heme biosynthesis and erythropoiesis. This report will summarize the EPA findings and will describe new studies.

The various steps in heme biosynthesis affected by lead are depicted in Figure 2-11. The activity of delta-aminolevulinic acid synthetase, ALA-S (the rate-limiting enzyme), is stimulated in various tissues in the presence of lead intoxication. Stimulation of enzyme activity begins to occur at a blood lead concentration of 40 μg/dL (Meredith et al., 1978). Accumulation of ALA in lead-exposed subjects occurs both by increased delta-ALA-S activity and by inhibition of activity of the extremely lead-sensitive cytosolic enzyme porphobilinogen synthetase (PBG-S, ALA dehydratase [ALA-D]). Of the two sources of ALA accumulation, that associated with ALA-D activity inhibition is predominant. The threshold for ALA-D inhibition is between 5 and 10 μg/dL. ALA accumulation in urine (ALA-U) has generally been assumed to begin in children and adults with blood lead concentration of about 40 μg/dL (e.g., NRC, 1972; EPA, 1986a). Buildup in plasma (ALA-P) can occur at lower blood lead concentrations (Meredith et al., 1978). Recent evidence (Okayama et al., 1989) suggests that the accumulation rate depends on the method of measurement of the metabolite; dose-response calculations of accumulation of ALA in lead workers suggest a threshold of below 20 μg/dL, on the basis of the more specific high-performance liquid chromatography.

Another important step affected by lead is formation of heme from protoporphyrin IX (erythrocyte protoporphyrin; zinc protoporphyrin) in which insertion of iron is inhibited by both inhibition of activity of the enzyme ferrochelatase (Piomelli, 1981; Posnett et al., 1988) and possibly altered transport of iron intramitochondrially (Moore, 1988; Marcus and Schwartz, 1987; Piomelli, 1981; Sassa et al., 1973). Lead-associated accumulation of zinc protoporphyrin (ZPP) resembles that due to iron deficiency in young children, and one must adjust for the presence of

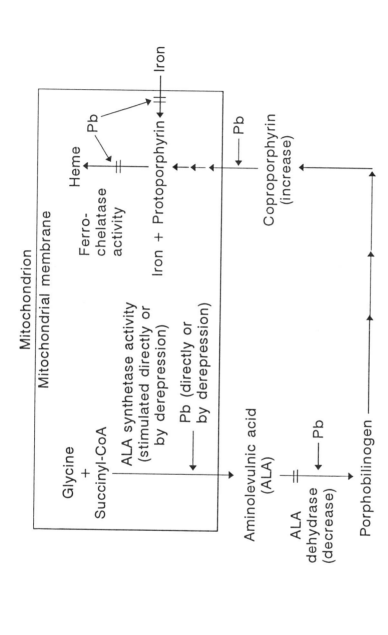

FIGURE 2-11 Schematic of mammalian heme synthesis pathway and steps impaired by lead. Source: Adapted from EPA, 1986a, Vol. IV.

iron deficiency when examining lead exposure. Accumulation of ZPP is strongly and logarithmically correlated with blood lead concentrations in both children (e.g., Piomelli et al., 1973, 1982; Lamola et al., 1975; Roels et al., 1976) and adults (e.g., Grandjean and Lintrup, 1978; Lilis et al., 1978). The threshold for response in children is 15-20 μg/dL (Piomelli et al., 1982; Hammond et al., 1985).

Lead-induced impairment of the body heme pool has a broad impact on the entire organism, and this can be readily comprehended as an effect-cascade scheme. In summary:

- Diminished hemoglobin biosynthesis leads eventually in a dose-dependent manner to anemia (EPA, 1986a; WHO, 1977; Schwartz et al., 1990). The threshold for lead's effect on hematocrit—based on analysis of a group of 579 children's blood data—is approximately 20-25 μg/dL (Schwartz et al., 1990).
- Heme formation for hemoproteins in neural tissues can be affected, disturbing normal brain-cell energy production (e.g., Whetsell et al., 1984; Holtzman et al., 1981).
- Lead disturbs heme-mediated formation of 1,25-$(OH)_2$-vitamin D in the kidney, and the disturbance can affect the many functions of vitamin D in calcium balance and metabolism (Edelstein et al., 1984; Mahaffey et al., 1982b; Fowler et al., 1980).
- Other steps in the heme pathway disturbed by lead include the accumulation of coproporphyrin in urine (CP-U) at an apparent blood-lead threshold of approximately 40 μg/dL, considerably above that for the steps already noted. This disturbance occurs only during lead intoxication (Piomelli and Graziano, 1980), in contrast to the delay observed in ZPP changes.

In general, impairments of erythropoiesis and erythrocyte physiology caused by lead exposure are considered to occur at relatively higher body burdens—i.e., blood lead at or above approximately 40 μg/dL—than is the case for effects on heme biosynthesis.

Iron-deficiency anemia in children is exacerbated by lead (Watson et al., 1980; Yip et al., 1981); lead workers' hemoglobin concentration is inversely and strongly correlated with blood lead concentrations at a threshold of approximately 50 μg/dL. In children, as noted earlier, lead exposure affects hematocrit at a blood-lead threshold of approximately 20-25 μg/dL (Schwartz et al., 1990).

Lead-associated anemia also occurs through direct damage to the erythrocyte, i.e., hemolytic anemia. This occurs through a combination of increased erythrocyte fragility (Waldron, 1966), increased osmotic resistance (e.g., Horiguchi et al., 1974), and impaired erythropoietic pyrimidine metabolism (Valentine and Paglia, 1980; Angle and McIntire, 1982).

Effects on Vitamin D and Calcium Metabolism

Recent studies document the impairment of vitamin D metabolism and function by lead, and this suggests a risk of a toxic action of lead on the many human bodily functions mediated by vitamin D, particularly during development. Those functions are tabulated in Table 2-3. Two studies of the toxic response in children have appeared in the literature (Rosen et al., 1980; Mahaffey et al., 1982b). In the report of Rosen et al. (1980), a group of lead-intoxicated children with blood lead concentrations between 33 and 120 μg/dL were observed to have had marked decreases in circulating amounts of the vitamin D metabolite 1,25-$(OH)_2$-vitamin D. This association was most pronounced, in a dose-dependent manner, at blood lead concentrations over 62 μg/dL, but reductions were still significant down to 33-55 μg/dL. The fact that chelation therapy in the children returned 1,25-$(OH)_2$-vitamin D to normal without affecting the vitamin D hydroxylation in the kidney indicated that activity of the mitochondrial enzyme P-450 in the kidney might have been impaired.

A second investigation in children also documented a strong negative correlation between blood lead concentrations and serum 1,25-$(OH)_2$-vitamin D concentrations (Mahaffey et al., 1982b). They found that the slopes of the regression analysis lines for data subsets above and below blood lead of 30 μg/dL were comparable for blood lead and 1,25-$(OH)_2$-vitamin D concentrations.

The decrements in circulating 1,25-$(OH)_2$-vitamin D seen by Rosen et al. (1980) in the blood lead range of 33-55 μg/dL are similar to those seen in children with severe renal insufficiency (Rosen and Chesney, 1983). For the entire blood lead range of 33-120 μg/dL, the metabolite reductions are similar to what is observed in vitamin D-dependent rickets (Type I), oxaloses, hormone-deficient hypoparathyroidism, and aluminum intoxication (Rosen and Chesney, 1983).

Data on experimental animals support an effect of lead on biosynthesis of 1,25-$(OH)_2$-vitamin D. Smith et al. (1981) observed decreases in the

TABLE 2-3 Various Tissues, Cell Types, and Functions Modulated by Vitamin D Hormone

Tissue	Cell Type	Function
Bone	Osteoblasts	Modulates cytosolic Ca^{2+} Bone remodeling
Gastrointestinal tract	Enterocyte	Mineral absorption
Mammary gland	Mammary explants	Calcium uptake
Parathyroid gland	Parathyroid	Phospholipid metabolism
Pituitary gland	GH_4C_1 pituitary cell	Prolactin synthesis Modulates cytosolic Ca^{2+}
Heart	Cardiac muscle	Calcium uptake
Skin	Fibroblasts, epidermal	CGMO production
Kidney	Tubular cells	Phosphate reabsorption
Skeletal muscle	Myoblasts	Calcium uptake
Pancreas	β cells	Insulin secretion
	Macrophage	Phagocytosis Modulation of proto-oncogenes

(continued, next page)

plasma metabolite concentrations in rats given high doses of lead orally. In chicks fed lead, a dose-dependent reduction in biosynthesis of the dihydroxy hormone in the kidney and reduced concentrations in tissues have been noted (Edelstein et al., 1984).

As can be seen in Table 2-3, 1,25-$(OH)_2$-vitamin D controls bone remodeling and several other metabolic functions. Additional recently

TABLE 2-3 (CONT.)

Tissue	Cell Type	Function
Diverse	Lymphocytes HL-60 myeloid leukemia cells ROS osteosarcoma cells Mononuclear cells Epidermal cells Cancer cell lines, multiple	Modulates differentiation and proliferation
Diverse	Diverse	Cell division Cell-cell communication Organization of cytoskeleton Endocrine secretion Neurotransmitter release Platelet release reaction Exocrine secretion

characterized functions of this vitamin D were summarized by Reichel et al. (1989). 1,25-$(OH)_2$-vitamin D, for example, controls intestinal vesicular calcium (Nemere and Norman, 1988) and modulates intracellular calcium ion in mouse osteoblasts (Lieberherr, 1987), rat cardiac muscle cells (Walters et al., 1987), and cultured mouse mammary gland (Mezzetti et al., 1988). Other test systems have also been investigated (Chisolm et al., 1988; Sugimoto et al., 1988).

Collectively, a decrease in amounts of the dihydroxy metabolite can potentially produce disturbances in the calcium messenger system and cell functions controlled by calcium ion (Pounds and Rosen, 1988).

Immunoregulatory properties of the dihydroxy metabolite have been noted (e.g., Iho et al., 1986), as has its role in growth and differentiation of cell types other than those of bone (see, e.g., Barsony and Marx, 1988). Other newly revealed functions are shown in Table 2-3.

A number of studies have attempted to measure the functions affected by lead, and these are considered in the section of this chapter dealing with molecular mechanisms of lead toxicity.

Carcinogensis

Virtually all the attention to lead as a major public-health problem arises from its noncarcinogenic effects in humans and experimental animals. But questions have been raised about lead carcinogenicity and the topic is briefly summarized here. Available data are from occupational epidemiologic studies, short- and long-term experimental-animal tests, and biochemical and in vitro assessments of lead compounds.

Various studies (Hiasa et al., 1983; Shirai et al., 1984; Tanner and Lipsky, 1984) have shown that dietary exposure to lead acetate at relatively high doses increases the development of renal cancer caused by several known organic renal carcinogens in rats. Hiasa et al. (1983) studied the promoting effects of a diet containing 1% lead acetate on the development of renal tubule-cell tumors after earlier exposure to N-ethyl-N-hydroxyl nitrosamine. They found a 70% incidence of renal tumors in animals given lead acetate and the nitrosamine at 32 weeks and no increase of tumors in animals given either compound alone. Similar results with the nitrosamine were reported by Shirai et al. (1984), who concluded that lead acetate acted as a promoter, and with N-(4'-fluoro-4-biphenyl)acetamide by Tanner and Lipsky (1984), who demonstrated that lead acetate accelerated the onset and development of kidney tumors in rats after chronic exposure.

Kasprzak et al., (1985) used a diet containing 1% lead acetate and supplemented with calcium acetate at 0-6% and showed that addition of calcium acetate to the diet tended to increase the incidence of renal tumors after 58 weeks of lead acetate exposure from 45% to 71%, but decreased the accumulation of lead in the kidneys.

The overall results indicate that lead can act both as a renal carcinogen in rodents and as a promoter of renal carcinogenesis caused by other organic renal carcinogens. The exact minimal doses of lead required to produce the effects are unknown.

About a dozen occupational studies have considered lead exposure versus various types of cancers in such work categories as battery recycling, lead smelting, alkyl lead manufacturing, plumbing, and pipefitting. EPA (1986a) has examined the older studies, and they have been augmented by those of Fanning (1988) for battery operations, Gerhardsson et al. (1986) for alkyl lead production, and Cantor et al. (1986) for plumbing and pipefitting.

Results of some studies suggest a renal-cancer risk (Selevan et al., 1985; Cantor et al., 1986), but results of others do not (e.g., Cooper et al., 1985). In some cases (Selevan et al., 1988), an association with duration of exposure added plausibility to the findings.

In contrast, results of numerous animal studies strongly support a renal-cancer potential for soluble lead salts (see EPA, 1986a; IARC, 1980, 1987), and at least 12 long-term studies (with various rat strains, a mouse strain, and both males and females) documented induction of renal tumors when animals were fed either lead acetate or subacetate (soluble forms of lead). Those animal data meet EPA's criteria for sufficient evidence of carcinogenicity, as published in "Guidelines for Carcinogen Risk Assessment" (EPA, 1986b).

The mechanism of lead carcinogenicity in laboratory animals remains unclear. Lead is not mutagenic in most test systems, but it has been shown to be clastogenic, reducing the fidelity of DNA repair polymerases. As described above, lead compounds are also mitogens in rodent kidneys and have been shown to act as tumor promoters and co-carcinogens in various experimental studies. The inconclusive human data and established animal carcinogenicity led to a classification of lead as a probable human carcinogen (IARC, 1980, 1987) and an EPA carcinogen ranking of Group B2 (EPA, 1986b). Since the U.S. population has an exposure of approximately 1 μg/kg of body weight per day, the EPA method for extrapolating animal data to humans could be used to estimate a lifetime cancer risk for lead of approximately 10^{-5}. Of course, the cancer risks would vary, depending on the extent of exposure if the linearized extrapolation is appropriate for lead based on biology. The noncarcinogenic effects of lead remain of predominant interest in sensitive populations, but the potential carcinogenic effects of lead should be considered as new information becomes available.

Nephropathy

The question arises whether lead has subclinical renal effects, as would be expected, given the array of toxic effects on other organ systems. Several obstacles frustrate efforts to answer the question epidemiologically or experimentally. First, there is a marked reserve capacity of the kidney to function in the face of toxic insult. It might be some time before the reserve capacity is depleted when people are exposed to large concentrations; this has been shown in workers occupationally exposed to lead. Second, we do not know the mechanisms of nephrotoxic events at the cellular or subcellular level, e.g., in the proximal renal tubules. Finally, there is a dearth of biologic markers specific for lead's nephrotoxic action.

Previous studies (Victery et al., 1984) have shown that lead-ion uptake in proximal renal tubule cells occurs via membrane binding or passive diffusion. Consequently, it is the kidney's intracellular handling of lead that defines the nephrotoxic dose-response relations for lead toxicity.

One can look to several kinds of experimental studies to garner clues as to what is occurring in humans who have subclinical lead exposures. Of particular interest are data on leadbinding proteins in experimental systems.

Oskarsson et al. (1982) showed that the lead-binding patterns in rat kidneys and brain, major target organs for lead toxicity, were consistent with binding to two proteins, which might thus be factors in the intracellular handling and availability of lead. Goering and Fowler (1984, 1985) showed that the inhibition of PBG-S (ALA-D) activity in the rat kidney is mediated by both lead biochelation and zinc availability; the former perhaps helps not only to account for relative resistance to lead in kidney PBG-S (Fowler et al., 1980; Oskarsson and Fowler, 1985a), a cytosolic enzyme otherwise quite sensitive to lead in other tissue (Fowler et al., 1980; Oskarsson and Fowler, 1985a), but such sequestration are linked to the presence of the lead binding proteins. ALA-S and ferrochelatase function were not, however, protected. These observations suggested that either other molecules or the lead-binding proteins were facilitating the mitochondrial uptake of lead, because the mitochondrial inner membrane has previously been shown (Oskarsson and Fowler, 1985a,b) to be highly impermeable to lead in vitro (Oskarsson and Fowler, 1987). Studies of Mistry et al. (1985) show a high affinity of these proteins for lead; other data (Fowler et al., 1985; Mistry et al., 1986; Shelton et al., 1986) indicate that they play a role in the intranuclear transport of lead and in lead-induced changes in renal gene expression and that their biologically active form is a cleavage fragment of alpha-2-microglobulin in the retinol-binding protein family (Fowler and DuVal, 1991) that undergoes aggregation, at least in vitro (DuVal and Fowler, 1989).

With experimental exposures to mercury (Woods and Fowler, 1977; Woods et al., 1984) or inorganic oxyarsenic (Woods and Fowler, 1978; Mahaffey et al., 1981), the resulting porphyrinuria appears to be derived from injury to the kidney itself. Other data are consistent with a lead-associated effect on renal heme formation. Ferrochelatase inhibition (a mechanism of erythrocyte protoporphyrin accumulation) occurs in the kidney (Fowler et al., 1980). The kidney is relatively rich in porphyrins (Zawirska and Medras, 1972; Maines and Kappas, 1977), and lead appears to inhibit the heme-requiring kidney 1-hydroxylase enzyme system (Rosen and Chesney, 1983; Reichel et al., 1989); one function of this system is the formation of 1,25-$(OH)_2$-vitamin D, the hormonal metabolite of vitamin D. Short-term experimental-animal studies with intravenous lead have shown a high correlation between formation and dissolution of lead inclusion bodies in renal tubule cells and changes in total and specific gene regulation at these sites. Such changes in regulation are to be found in various subcellular fractions—mitochondrial, microsomal, cytosolic, and lysosomal fractions with a specific response for each organelle compartment (Mistry et al., 1987).

Chronic exposures in experimental animals have produced similar results with regard to lead induction of specific stress proteins. Comparisons of such data with morphometric analyses of tubule cell populations and effects on heme biosynthesis might permit determination of which biologic markers are of greater utility in delineating specific lead-induced changes in cell functions.

Knowledge of the intracellular handling of lead in target tissues, such as the kidney and brain, is essential to an understanding of the mechanisms of lead toxicity in target cell populations in these tissues. Soluble, high-affinity lead-binding proteins in the kidney and brain of rats were first reported by Oskarsson et al. (1982). Those molecules were not identified in other nontarget tissues, so they might play a role in lead toxicity in these organs at low doses. Later studies (Goering and Fowler, 1984, 1985; Goering et al., 1986) demonstrated that semipurified preparations of the proteins play a major role in mediating lead inhibition of the heme biosynthetic pathway enzyme alpha-aminolevulinic acid dehydratase (porphobilinogen synthetase). Other studies (Mistry et al., 1985, 1986) demonstrated that the kidney lead-binding proteins had a high affinity for lead and were capable of facilitating the cell-free nuclear translocation and chromatin binding of ^{203}Pb. Those molecules thus appear to act as "receptors" for lead and to regulate its intranuclear uptake, chromatin binding, and changes in proximal tubule cell gene expression (Fowler et al., 1985; Shelton et al., 1986; Mistry et al., 1987; Hitzfeld et al., 1989; Klann and Shelton, 1989).

More recent studies (DuVal et al., 1989; Fowler and DuVal, 1991) have identified the renal lead-binding protein as a cleaved form of the protein alpha-2-microglobulin that locks the first 9-nitrogen-terminal residue and shown that the brain lead-binding protein is a chemically similar protein that is rich in aspartic and glutamic amino acids, but immunologically distinct. It appears that it is the cleared form of the alpha-2-microglobulin that is biologically active. Western blot studies (DuVal et al., 1989) have shown that that protein undergoes aggregation after in vitro exposure to lead. The data suggest that the protein can play an early role in the formation of the pathognomonic cytoplasmic and intranuclear lead inclusion bodies. The inclusion bodies are the main intracellular storage sites for lead in proximal tubule cells after increased or chronic lead exposure (Goyer and Rhyne, 1973; Moore et al., 1973; Fowler et al., 1980; Shelton and Egle 1982; Oskarsson and Fowler 1985a,b; Klann and Shelton 1989).

Previous studies have shown a marked kidney-specific macromolecular binding pattern for lead in renal proximal tubule cells. The binding is followed by several undefined intracellular events that result in the presence of large quantities of lead in renal proximal tubule cell nuclei. The precise sequence of events and relationships to other intracellular lead species have not been completely studied. Such data are central to an understanding of the bioavailability of lead to sensitive cellular processes. The bioactive lead species thus might be centrally involved in the mechanisms of toxicity. Further studies of how lead reacts with them should permit their use as biologic indicators of lead-induced nephropathy, provided that they are excreted in urine. Lead-induced alterations of renal gene expression (Fowler et al., 1985; Mistry et al., 1986; Shelton et al., 1986; Herberson et al., 1987; Hitzfeld et al., 1989) and inhibition of renal heme biosynthetic pathway enzymes with attendant development of metal-specific porphyrinuria patterns (Mahaffey and Fowler, 1977; Mahaffey et al., 1981) are examples of sensitive biochemical systems. Such responses have great potential as biologic indicators of nephropathy, once their relation to the pathophysiology of lead nephropathy is understood.

SUMMARY

Exposure of various sensitive populations to lead induces a wide variety of adverse effects—in the central nervous system of children and fetuses, in various growth indexes of children, in the cardiovascular system of older people, in heme synthesis, and in calcium homeostasis and function. LOELs (lowest-observed-effect levels) for various lead effects are summarized in Table 2-4 for children and Table 2-5 for adults.

The weight of the evidence gathered during the 1980s clearly supports the conclusion that the central and peripheral nervous systems of both children and adults are demonstrably affected by lead at exposures formerly thought to be well within the safe range. In children, blood lead concentrations around 10 μg/dL are associated with disturbances in early physical and mental growth and in later intellectual functioning and academic achievement. Studies of electrophysiologic end points have suggested some of the changes in brain function that might mediate the apparent effects.

Despite impressive advances over the last decade in the methodologic rigor of studies of lead exposure and nervous system function, epidemiology remains limited by opportunity. Therefore, animal studies are critical for interpreting the human data. Factors that an epidemiologist must take account of with statistical analysis (an inevitably imperfect process) can be controlled experimentally with animals. For example, the influence of socioeconomic factors on performance can be eliminated, and the importance of timing, dose, and duration of exposure can be evaluated more precisely.

The extensive evidence gathered in animal studies cannot be reviewed here, but some themes warrant enumeration. First, primates exposed to sufficient lead to produce a blood lead concentration of 25 μg/dL or less manifest a variety of memory, learning, and attentional deficits resembling those observed in humans. Second, the deficits appear to be permanent; they are evident for as long as 10 years in animals whose blood lead is maintained at approximately 15 μg/dL. Third, striking concordance of the human and animal data weighs heavily in favor of the hypothesis that low-dose lead exposure is responsible for some of the developmental and cognitive deficits observed in humans. Many of the

TABLE 2-4 Lowest-Observed-Effect Levels of Blood Lead for Effects in Children

LOEL, μg/dL	Neurologic Effects	Heme-Synthesis Effects	Other Effects
<10 to 15 (prenatal and postnatal)	Deficits in neurobehavioral development (Bayley and McCarthy Scales), electro-physiologic changes,[a,b] and lower IQ[c,d]	ALA-D inhibition[e]	Reduced gestational age and birthweight; reduced size up to age 7-8 yr[a,b,e]
15-20		Erythrocyte protoporphyrin increase[a,e]	Impaired vitamin D metabolism, Py-5'-N inhibition[a,e]
<25	Longer reaction time (studied cross-sectionally)[b,e]	Reduced hematocrit (reduced Hb)[f]	
30	Slower nerve conduction[e]		
40		Increasing CP-U and ALA-U[c]	
70	Peripheral neuropathies[a,e]	Frank anemia[a,e]	
80-100	Encephalopathy[a,e]		Colic, other gastrointestinal effects, kidney effects[e]

[a]Data from CDC, 1991.
[b]Data from EPA, 1990a,b.
[c]Data from Bellinger et al., 1992.
[d]Data from Dietrich et al., 1993a.
[e]Data from ATSDR, 1988.
[f]Data from Schwartz et al., 1990.

TABLE 2-5 Lowest-Observed-Effect Levels of Blood Lead for Effects in Adults

LOEL, μg/dL	Heme Synthesis and Hematologic Effects	Neurologic Effects	Renal Effects	Reproductive Effects	Cardiovascular Effects
<10	ALA-D inhibition				
10-15					Increased blood pressure
15-20	Erythrocyte protoporphyrin increase in females				
25-30	Erythrocyte protoporphyrin increase in males				
40	Increased ALA-U and CP-U	Peripheral nerve dysfunction (slower nerve conduction)			
50	Reduced hemoglobin production	Overt subencephalopathic neurologic symptoms		Altered testicular function	

LOEL, μg/dL	Heme Synthesis and Hematologic Effects	Neurologic Effects	Renal Effects	Reproductive Effects	Cardiovascular Effects
60				Female reproductive effects	
80	Frank anemia				
100-120		Encephalopathic signs and symptoms	Chronic nephropathy		

Source: Adapted from EPA, 1986a, Vol. IV.

neurodevelopmental and possibly other toxic effects resulting from lead exposure might not be reversible. It is important to distinguish two aspects of reversibility. The first pertains to biologic plasticity, specifically an organism's ability to repair damage and recover functional capacity. The second pertains to the reality of exposure patterns. Impairment might persist, regardless of an organism's capacity for recovery, as long as exposure is maintained. In a practical sense, the impairment is irreversible. Therefore, the fact that an adverse effect is reversible in the biologic sense does not necessarily mean that it is without potential public-health importance. The key issue is whether exposure is reduced, and expression of an organism's recovery capacities thus permitted.

This chapter documents that lead induces measurable increases in diastolic and systolic blood pressure in human populations and in experimental-animal models of environmentally induced blood-pressure changes.

Lead exposure is not the only risk factor for hypertension, but is more amenable to reduction or prevention than behavioral factors that are refractory to change. Furthermore, the relation of lead to blood pressure persists across a dose-effect continuum, so reducing lead exposure of all magnitudes has public-health and societal ramifications. Lead's impact is noteworthy also because of the importance of associated cardiovascular morbidity and mortality, even for an agent that contributes less than a major risk.

Through various processes, including toxicokinetic and intracellular disturbances, lead impairs calcium homeostasis and functions. The importance of impacts on calcium is that they impair calcium's central role in multiple cellular processes.

Lead produces a cascade of effects on the heme body pool and affects heme synthesis. Some of the effects serve as early measures of body lead burden.

Lead effects on cognition and other neurobehavioral measures need to be evaluated on a population-wide, as well as individual, basis. This evaluation should account for the whole statistical distribution of exposures and associated toxicity.

Unfortunately, the direct identification and linkage of the critical and sensitive biological processes which are targets for these effects remains saltatory. There are many reasons why our ability to define the mechanism(s) of action for lead toxicity lags behind our ability to detect and quantify the toxicological effects. In addition to the difficulties in defining a mechanism of action as discussed above, these reasons include:

1. Lead is a catholic toxicant producing adverse effects in most tissues and organs of the body, with a parallel effects on multiple organelles and metabolic processes. This situation makes it extremely difficult to identify and isolate the critical process(es) with sufficient experimental rigor.

2. There is frequently a long delay between the onset of lead exposure and the development of toxic manifestations, impairing identification of causal relationships between functional and cellular or biochemical events.

3. Lead causes nonspecific, decremental loss of tissue and organ function, with no important pathognomonic manifestations of toxicity.

4. The multifactorial nature of the toxicity in the nervous, cardiovascular, skeletal and other organ systems complicate establishing causal relationships between cellular and molecular processes and organ dysfunction.

Nevertheless, these difficulties do not diminish the importance or necessity of these efforts. Continued efforts must be made to bridge experimental animal and human studies, at all level of analysis, and to integrate the biochemical and molecular events impacting function at the level of the whole organisms.

3

Lead Exposure of Sensitive Populations

A complete assessment of exposure in sensitive populations requires knowledge of the sources of exposure. That is especially important for lead: it has multiple sources, and knowledge of them helps to define exposure to lead and to identify sensitive populations.

The conventional approach to identifying lead exposure in a population has been to attribute lead intoxication to single sources of lead at high concentrations, such as leaded paint. However, current understanding calls for a more comprehensive view. First, there is a growing consensus that lead induces a continuum of toxic effects in humans, starting with small exposures that cause subtle, but important, early effects. Our understanding of what constitutes a safe exposure has increased; as a result, the upper limit of a safe lead content in blood has declined to one-sixth to one-fourth of what it was in a matter of a few decades. Second, once lead is absorbed from a specific source, it is added to a body burden that contributes to various health effects. Therefore, exposures small enough to have been viewed as of little importance now are taken more seriously. In other words, we must consider the aggregate impact of multiple small lead sources in assessing health risk.

This chapter is divided into three sections. The first provides a historical perspective on lead contamination, addressing such topics as natural concentrations of environmental lead and the chronologic record of anthropogenic contamination with lead. The second section discusses the major current sources and pathways of lead exposure in sensitive populations, including paint, air, dust and soil, and drinking water and food. The section includes a brief discussion of occupational lead exposure and ends with sources that can produce large, but not necessarily pervasive, exposure, such as improperly lead-glazed food and beverage containers and lead-based ethnic medicinal preparations. The chapter concludes with a detailed summary.

HISTORICAL OVERVIEW OF ANTHROPOGENIC LEAD CONTAMINATION

Lead production dates to the discovery of cupellation—a metallurgic process for separating silver from lead ores—some 5,000 years ago (Nriagu, 1985a). However, such anthropologic artifacts as the lead beads in the Hittite ruins of Catal Hüyück from 6500 BC and the lead statuette from the temple of Osiris in Abydos from 3000 BC reveal earlier uses of lead.

The historical record of industrial lead production over the last 5,000 years is illustrated in Figure 3-1. The current production rate is approximately 3.4 million metric tons per year (U.S. Bureau of Mines, 1989). The total amount of lead mined over the last 5,000 years is estimated to be 300 million metric tons (Flegal and Smith, 1992).

Lead has a long history of wide use. A lead glaze in a Babylonian tablet from 1700 BC has been described; these glazes had become common in China during the Chou Dynasty of 1122-256 BC. In the Roman Empire, lead was used in cooking pots and other utensils, in syrups, in beverage adulterants (e.g., sapa), in medicines, and in the construction of pipes and cisterns to transport water (Nriagu, 1983b). The wide use of lead for the latter explains the word plumbing (from the Latin *plumbum*, lead). Lead was so pervasive during that period that there is little doubt that lead poisoning was endemic in the Roman population. In fact, it has been speculated (Gilfillan, 1965; Nriagu, 1983b) that chronic lead poisoning contributed substantially to the decline of the Roman Empire.

One of the environmental tragedies of that period is that, despite some Romans' recognition of the societal problems associated with lead toxicity, awareness did not restrict its use. For example, Vitruvius (Nriagu, 1985b) observed:

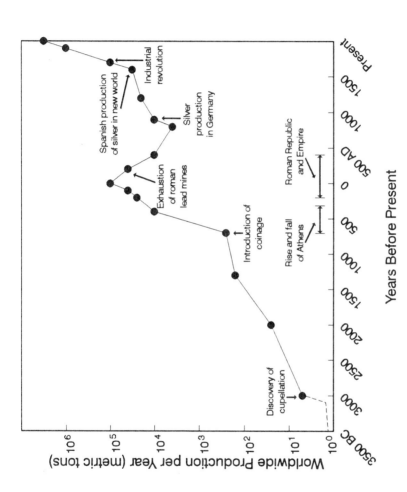

FIGURE 3-1 Historical record of industrial lead production in last 5,000 years. Source: Adapted from Settle and Patterson, 1980.

Water supply by earthen pipes has advantages. First, if any fault occurs in the work, anybody can repair it. Again, water is much more wholesome from earthenware pipes than from lead pipes, for it seems to be made injurious by lead because some white lead is produced from it; and this is said to be harmful to the human body. Thus if what is produced by anything is injurious, it is not doubtful but that the thing is not wholesome in itself.

We can take example by workers in lead who have complexions affected by pallor. For when, in casting, the lead receives the current of air, the fumes from it occupy the members of the body, and burning then thereupon, robs the limbs of the virtues of the blood. Therefore, it seems that water should not be brought in lead pipes if we desire to have it wholesome.

Current uses of lead are much more extensive. It is still used in some glazes, eating utensils, folk medicines, and plumbing. It is also used in paint pigments, solders, wall and window construction, cosmetics, sheeting of ships, roofs, guttering, containers, sealants, protective coatings, printing type, insecticides, batteries, plastics, lubricants, ceramics, machine alloys, and gasoline additives (NRC, 1980; EPA, 1986a).

The amount of contaminating lead released into the environment closely parallels the record of lead production over the last 5,000 years. Approximately half the lead produced is released into the environment as contamination (NRC, 1980). Current production is about 3.4 million metric tons per year, and current lead release is about 1.6 million metric tons per year. About 150 million metric tons of lead has been released into the environment in the last 5,000 years. The latter value, total release, is probably closer to the total amount of lead put to use, approximately 300 million metric tons, inasmuch as the element is indestructible and cannot be transformed into an innocuous form.

Much of the lead released into the environment is emitted into the atmosphere (about 330,000 metric tons/year) (Nriagu and Pacyna, 1988). Those releases are currently dominated by emissions from leaded gasoline (over 248,000 metric tons/year), but emissions from other sources—including coal and oil combustion, mining, manufacturing, incineration, fertilizers, cement production, and wood combustion—are substantial (Table 3-1). In fact, the latter exceed emissions of most other contaminants by orders of magnitude.

The magnitude of industrial emissions of lead is illustrated by comparisons with natural emissions of lead and other contaminants. The

TABLE 3-1 Worldwide Emissions of Lead to the Environment, 1983[a]

Source	Amount, 10^3 kg/yr
Coal combustion	1,765-14,550
Oil combustion	948-3,890
Mining	30,060-69,640
Manufacturing	1,065-14,200
Incineration	1,640-3,100
Fertilizers	55-274
Cement production	18-14,240
Wood combustion	1,200-3,000
Leaded gasoline	248,030
Miscellaneous	3,900-5,100
Total	288,700-376,000

[a]Data from Nriagu and Pacyna, 1988.

sum of industrial lead emissions is approximately 700 times the sum of natural emissions of lead into the atmosphere (Patterson and Settle, 1987; Nriagu, 1989). Emission of industrial lead aerosols to land and aquatic ecosystems is now predominant. It accounts for approximately 15-20% (202,000-263,000 metric tons/year) of the total anthropogenic emission of lead to land (approximately 1,350,000 metric tons/year) and approximately 63-82% (87,000-113,000 metric tons/year) of the total lead that enters aquatic ecosystems (approximately 138,000 metric tons/year) (Nriagu and Pacyna, 1988).

The historical record of atmospheric emissions of industrial lead aerosols has been measured in the environment by various investigators (Figure 3-2). It was initially documented by the 230-fold increase in lead deposition rates in Greenland ice cores over the last 3,000 years, from 0.03 ng/cm^2 per year in prehistoric ice cores (800 BC) to about 7 ng/cm^2 per year in contemporary ice cores (Murozumi et al., 1969; Ng and Patterson, 1981; Wolff and Peel, 1985). Comparable increases in the Northern Hemisphere have since been documented in pond and lake

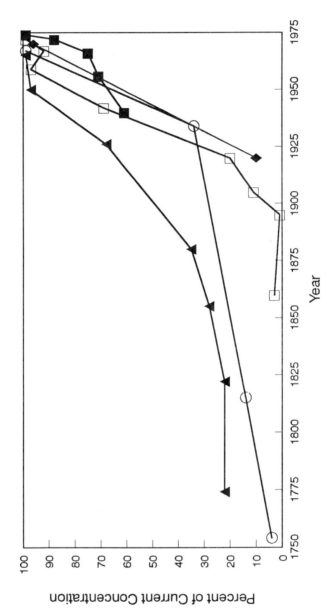

FIGURE 3-2 Lead contamination from industrial aerosols as recorded in chronologic strata. Circles, Greenland snow (Murozumi et al., 1969); squares, dated pond sediment from remote Sierras (Shirahata et al., 1980); open triangles, lake sediments (Edgington and Robbins, 1976); closed triangles, marine sediments (Ng and Patterson, 1982). Source: Adapted from EPA, 1986a, Vol. II.

sediments (Lee and Tallis, 1973; Edgington and Robbins, 1976; Robbins, 1978; Livett et al., 1979; Shirahata et al., 1980; Davis et al., 1982) the oceans (Schaule and Patterson, 1981, 1983; Flegal and Patterson, 1983; Boyle et al., 1986), pelagic sediments (Veron et al., 1987; Hamelin et al., 1988), and marine corals (Shen and Boyle, 1988).

Smaller increases by a factor of 2-5 have been detected in Antarctic ice cores (Boutron and Patterson, 1983, 1986; Patterson et al., 1987) and in the South Pacific (Flegal and Patterson, 1983; Flegal, 1986). The contrast reflects the localization of 90% of lead emissions in the northern hemisphere and the short residence time (10 days) of lead aerosols relative to the interhemispheric mixing rate of 1-2 years (Turekian, 1977; Flegal and Patterson, 1983).

Other releases of lead to the land range from 540,000 to 1,700,000 metric tons/year (Nriagu and Pacyna, 1988). These include industrial lead from commercial wastes, smelter wastes, and mine tailings (each approximately 300,000 metric tons/year); fly ash (approximately 140,000 metric tons/year); urban refuse (approximately 40,000 metric tons/year); agricultural wastes (approximately 14,000 metric tons/year); animal wastes (approximately 12,000 metric tons/year); solid wastes (approximately 8,000 metric tons/year); wood wastes (approximately 7,000 metric tons/year); municipal sewage sludge (approximately 6,000 metric tons/year); peat (approximately 2,000 metric tons/year); and fertilizers (approximately 1,000 metric tons/year). Many of those are projected to increase and become, at least relatively, more important with the reduction in atmospheric emission of gasoline lead.

Nonatmospheric input of industrial lead into aquatic ecosystems is smaller, but still substantial (Nriagu and Pacyna, 1988). It ranges from 25,000 to 50,000 metric tons/year and includes lead from manufacturing (approximately 14,000 metric tons/year), sewage sludge (approximately 9,000 metric tons/year), domestic wastewater (approximately 7,000 metric tons/year), smelting and refining (approximately 6,000 metric tons/year), and mining (1,000 metric tons/year).

Lead contamination in urban areas is often much greater than in remote areas (Table 3-2). That is due to the extensive use of lead in industrial processes and the relatively limited mobility of a sizable fraction of this lead. Long-distance transport of a fraction of the lead to the atmosphere also occurs. Terrestrial, aeolian, and fluvial gradients show that most of the lead emitted in urban areas has remained as a

TABLE 3-2 Environmental Lead Concentrations in Remote and Rural Areas and Urban Areas[a]

	Remote and Rural Lead Concentration, $\mu g/g$[b]	Reference	Urban Lead Concentration, $\mu g/g$[b]	References
Air	0.05	Lindberg and Harriss, 1981	0.3	Facchetti and Geiss, 1982; Galloway et al., 1982
Fresh water	1.7×10^{-5}	Elias et al., 1982	0.005-0.030	EPA, 1986a, Vol. II
Soil	10-30	EPA, 1986a, Vol. II	150-300	EPA, 1986a, Vol. II
Plants	0.18[c]	Elias et al., 1982	950[d]	Graham and Kalman, 1974
Herbivores (bone)	2.0[d]	Elias et al., 1982	38[d]	Chmiel and Harrison, 1981
Omnivores (bone)	1.3[d]	Elias et al., 1982	67[d]	Chmiel and Harrison, 1981
Carnivores (bone)	1.4[d]	Elias et al., 1982	193[d]	Chmiel and Harrison, 1981

[a]Values can be highly variable, depending on organism and habitat location.
[b]Except $\mu g/m^3$ in air.
[c]Fresh weight.
[d]Dry weight.

contaminant in those areas (Huntzicker et al., 1975; Roberts, 1975; Ragaini et al., 1977; Biggins and Harrison, 1979; Palmer and Kucera, 1980; Harrison and Williams, 1982; Ng and Patterson, 1982; Elbaz-Poulichet et al., 1984; Flegal et al., 1989). For example, in the Great Lakes (Flegal et al., 1989), surface-water lead concentrations in the highly industrialized Hamilton harbor (290 pmol/kg) are nearly 50 times higher than those of some offshore waters in Lake Ontario (6.5 pmol/kg). Complementary stable lead-isotope composition measurements show that essentially all (over 99%) of that lead, in even the most remote regions of Lake Ontario and Lake Erie, is derived from releases of industrial lead from Canada and the United States.

Those measurements are consistent with those in numerous other studies that have shown the pandemic scale of lead contamination, which has increased lead concentrations throughout the Northern Hemisphere by a factor of at least 10. Lead concentrations in the atmosphere are now 100 times natural concentrations (Patterson and Settle, 1987). Lead concentrations in remote surface waters of the North Pacific and the North Atlantic are at least 10 times natural concentrations (Flegal and Patterson, 1983; Boyle et al., 1986). Lead concentrations in terrestrial organisms are 100 times natural concentrations (Elias et al., 1982).

Studies incorporating rigorous trace-metal analysis have shown that the natural background lead concentration of North American Indians in pre-Columbian times was 0.3 mg per 70-kg adult (Patterson et al., 1987; Ericson et al., 1991). The body of an average North American urban adult contains 100-1,000 times as much lead.

Some uses of lead are being reduced in the United States and other countries in response to growing concern over pervasive lead toxicity even at low exposures. For example, lead in gasoline has been decreased in recent decades (Figure 3-3), as noted widely (EPA, 1986a; Nriagu, 1990). The United States has also seen a major reduction in the use of lead-soldered cans for foods and beverages (EPA, 1986a; ATSDR, 1988); lead in such containers can increase food lead content by a factor of up to 4,000 over the lead content of fresh food (Settle and Patterson, 1980).

The dispersion of industrial lead is not constrained by national boundaries. For example, stable-isotope composition measurements, which can identify specific sources of industrial lead, have shown that industrial lead from Canada and the United States is transported across the

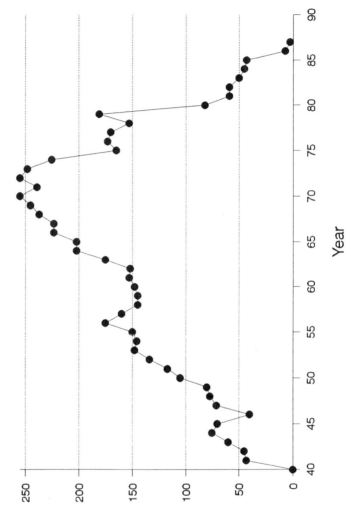

FIGURE 3-3 Lead in gasoline in United States. Source: Adapted from Nriagu, 1990.

Great Lakes (Flegal et al., 1989). Similar analyses have documented that over 95% of the lead in the North Pacific represents deposition of Asian and North American industrial lead aerosols (Figure 3-4).

SOURCE-SPECIFIC LEAD EXPOSURE OF SENSITIVE POPULATIONS

This chapter presents a general picture of the common modes of human exposure to lead—through leaded paint, air (which it enters from leaded gasoline and stationary emission), dust and soil, tap water, the workplace, and miscellaneous sources. Many of the sources and pathways of lead exposure are connected in ways that complicate exposure analysis and frustrate reduction and removal strategies (Figure 3-5); in this regard, lead in the air, lead in paint, and lead in drinking water are of particular concern.

Lead in Paint

Lead-based paint in and around U.S. urban housing has long been recognized as a serious and pervasive source of lead poisoning of young children. It also accounts for exposure to lead through its appearance in dust and soils. This source of lead poisoning has expanded to include workers in housing-lead abatement and homeowners who attempt rehabilitation of old housing. It also affects such workers as salvagers, construction crews, and marine maintenance staff who encounter mobilized lead in burning, cutting, chipping, and grinding.

Physicochemical and Environmental Considerations

Lead compounds have served as pigments for painting media for millennia; for example, the use of white lead pigment—basic lead carbonate—dates to prehistoric times (Friedstein, 1981). Older leaded paints included a linseed-oil vehicle plus a lead-based pigment and in some cases a long-chain fatty acid and a lead-based drying catalyst, or

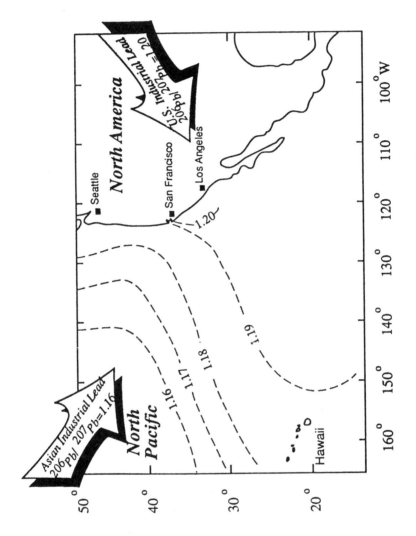

FIGURE 3-4 Movement of industrial lead aerosols to northeast Pacific Ocean from Asian and North American sources. Source: Adapted from Flegal and Stukas, 1987.

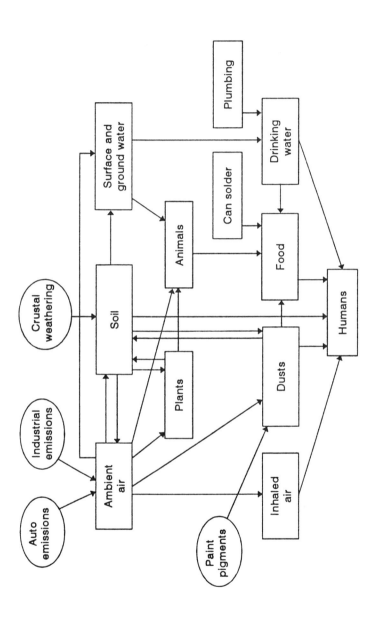

FIGURE 3-5 Sources and pathways of lead from environment to humans. Source: Adapted from EPA, 1986a, Vol. II.

drier, commonly an organic acid salt of lead. Use as pigment accounted for most of the lead present in older paints. Pigment lead concentrations were high in paints marketed and used before the 1940s; fractions in the final dry film were up to approximately 50% (CDC, 1985).

Physical properties of lead-based paints were such as to lead to their widespread use in homes, in public facilities, and in industrial sites. The most common pigment, basic lead carbonate ($2PbCO_3Pb(OH)_2$), reacted with other paint components to yield a flexible, durable lead soap film. The surface would have excellent durability (weathering) characteristics, but would eventually age, i.e., would either peel or undergo weathering and interior shedding or chalking. Aging yielded paint chips and a constantly renewing surface with concomitant dispersion of the older leaded film as a leaded dust on nearby surfaces.

The period during which leaded paint had the highest amounts of lead and posed the worst toxicity risk in the United States was from around 1875 to the 1940s, when other pigments, such as titanium dioxide, began entering the paint market (e.g., Farfel, 1985). Lead was used in residential paints throughout the 1800s, and it has been estimated (ATSDR, 1988) that 3 million metric tons of lead in various old paints persist in old housing and public facilities in the United States. Lead has been used in other pigments, e.g., as lead oxide and lead chromate.

General Characteristics of Exposure

Given the pervasive nature of leaded paint in homes, elementary schools, day-care centers, etc., and the normal oral exploratory behavior of very young children, it is logical for leaded paint to be a major source of lead for young children. Young children, especially toddlers, can ingest fallen or peelable chips of leaded paint, gnaw intact lead-painted woodwork, and ingest leaded paint dispersed in soils or in dusts adhering to hands. Household-paint dust can also be entrained into the breathing zone of toddlers and inhaled.

Scope of the Problem

In the United States, the distribution of paint lead in housing is a

TABLE 3-3 Estimated Numbers of Children Under 7 Years Old Residing in Lead-Based-Paint U.S. Housing, by Date of Construction

Construction Date	No. Lead-Based-Paint Homes	No. Children
Pre-1940	20,505,000	5,885,000
1940-1959	16,141,000	4,632,000
1960-1974	5,318,000	1,526,000
Total pre-1975	41,964,000	12,043,000

Source: Adapted from ATSDR, 1988. Data from Pope, 1986.

function of housing age, as shown in Table 3-3 (ATSDR, 1988), which also shows numbers of children in housing of different ages (derived from U.S. Bureau of the Census enumerations). In Table 3-3, the national estimate for the number of all housing units having paint with lead at or above the detection minimum of 0.7 mg/cm^2 is approximately 42 million, about 52% of the entire U.S. housing inventory. Of the 42 million, approximately 21 million units were built before 1940.

The number of children under 7 years old in lead-based-paint housing is about 12 million, of whom approximately 6 million live in the oldest units, which have the highest concentrations of lead in paint. It has been determined (ATSDR, 1988) that 4.4 million metropolitan children (children in 318 U.S. Standard Metropolitan Statistical Areas) 0.5-5 years old live in the oldest housing.

The percentages of housing with leaded paint by date of construction are pre-1940, 99%; 1940-1959, 70%; and 1959-1974, 20% (Pope, 1986). The oldest housing group also had the highest percentages of lead in paint formulations; percentages declined afterward (EPA, 1986a; ATSDR, 1988). The oldest U.S. housing is to be found in the older areas of the nation. Figure 3-6 shows that the northeastern and midwestern areas of the nation have the highest percentages of pre-1940 housing.

In 1989-1990, the U.S. Department of Housing and Urban Development (HUD) conducted a survey to estimate better the extent of the lead-based-paint hazard in the U.S. housing stock (HUD, 1990). Among the 77 million privately owned and occupied homes in the

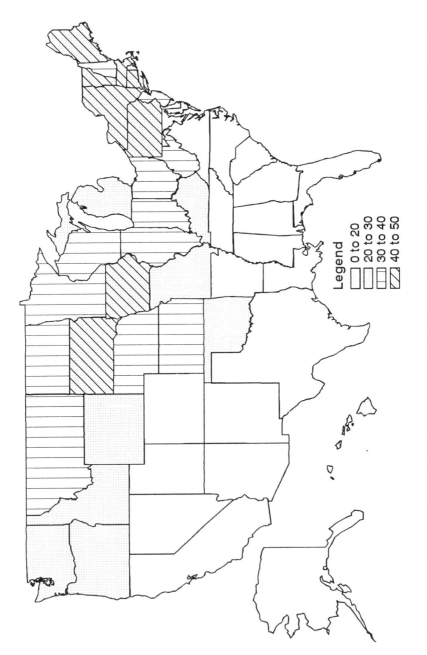

FIGURE 3-6 Percentage of U.S. housing constructed before 1940. Source: U.S. Bureau of the Census, 1986.

United States built before 1980, 57 million contained lead-based paint (defined as paint lead concentrations of at least 1.0 mg/cm^2). Families with children under 7 years old occupied an estimated 9.9 million of the 57 million; the 9.9 million included 3.7 million units with deteriorating (e.g., peeling) lead-based paint. The HUD survey provided additional detail on the location of the lead-based paint. Of the 57 million units with lead-based paint, 18 million had the paint only on exterior surfaces, 11 million only on interior surfaces, and 28 million on both.

Pope (1986) also determined (Table 3-4) from U.S. Bureau of the Census (1986) housing-survey data that about 6.2 million U.S. housing units are deteriorating and have leaded paint in unacceptable amounts. (As seen in Table 3-4, almost 1 million of these units were of pre-1940 construction.) Some 1.8 million children under 7 years old live in those units. Of these, 1.2 million are estimated by ATSDR (1988) to have blood lead concentrations over 15 μg/dL.

- *Case reports and case series.* As noted in reports from ATSDR (1988) and NRC (1972), the clinical literature of the last 60 years is full of case reports and reviews documenting severe lead poisoning of young children—laboratory evidence of lead in blood, paint chips in the gastrointestinal tract, and no concurrent environmental evidence of other sources of lead exposure.

- *Field epidemiology.* A particularly comprehensive data set, quantitatively linking child screening populations to leaded paint, is that from the 1976-1980 screening of children for lead poisoning in Chicago and the accompanying assessment of 80,000 housing units for the presence of leaded paint (Annest and Mahaffey, 1984). Schwartz and Levin (1991) analyzed the data and estimated a relative risk of approximately 15 for lead toxicity in summer months for children who resided in homes with leaded paint.

- *Research epidemiology.* Numerous site-specific epidemiologic studies have been critically evaluated (e.g., EPA, 1986a). They have entailed multivariate regression analyses in which the size of the paint lead contributions to blood lead concentrations are calculated. A particularly detailed study is that of children in inner-city housing in Cincinnati, Ohio, with leaded paint and paint-related pathways of exposure (Clark et al., 1985; Bornschein et al., 1987). Figure 3-7 shows data from Clark et al. (1985, 1987) as reanalyzed by this com-

TABLE 3-4 Estimated Numbers of Children Under 7 Years Old Residing in Unsound and Lead-Based-Paint U.S. Housing, by Age and Criteria of Deterioration

Category of Unsoundness	Construction Date	No. Unsound Lead-Based-Paint Homes	No. Children
Peeling paint	Pre-1940	964,000	277,000
	1940-1959	758,000	218,000
	1960-1974	250,000	72,000
Total peeling paint	Pre-1975	1,972,000	567,000
Broken plaster	Pre-1975	1,594,000	458,000
Holes in walls	Pre-1975	2,602,000	747,000
Grand totals	Pre-1975	6,168,000	1,772,000

Source: Adapted from ATSDR, 1988. Data from Pope (1986) and 1983 housing survey data (U.S. Bureau of the Census, 1986).

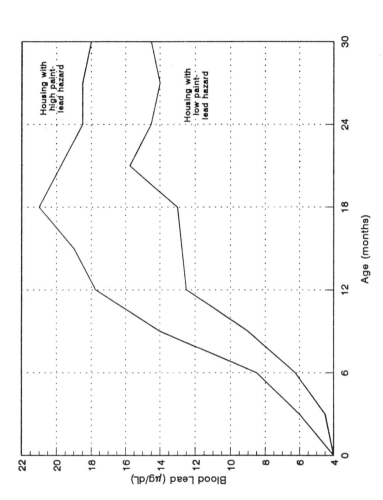

FIGURE 3-7 Longitudinal blood lead profiles of groups of children born and raised in housing with and without high-lead paint. Both classes of housing surrounded by lead-contaminated soil and dust. Data from Clark et al., 1985, 1987.

mittee; blood lead concentration is seen to change as a function of paint lead in housing.

HUD (1990) cited several published and unpublished studies on the influence of home renovation or abatement of lead-based paint on the blood lead concentration of children living in housing units during these activities. Bellinger et al. (1986b) reported a significant association between blood lead concentrations at age 24 months and recent home-refinishing activities. Rabinowitz et al. (1985a) reported a mean increase in blood lead concentrations (1.4 μg/dL, standard error 0.7) in children in homes recently refinished. Farfel (1987) reported that neither traditional nor modified methods of abating lead-based paint reduced the blood lead concentrations of children living in the residences. Farfel also reported that, at least in the short term, traditional abatement methods resulted in increased blood lead concentrations, presumably because of exposure to lead-laden dust. In contrast, three unpublished studies cited in the 1990 HUD report demonstrated blood lead reductions after traditional abatement methods.

Lead in Air

Lead enters air from gasoline and from stationary emissions. Environmental lead contamination from combustion of leaded gasoline has been widely documented in the United States and elsewhere (NRC, 1980; EPA, 1986a), and there is much evidence that it has added substantially to the body lead burdens of affected human populations.

From the 1920s to the late 1980s, lead was added to gasoline in the antiknock additive tetraethyl lead (later, this was mixed with tetramethyl lead). Tetraethyl lead is still used widely as a gasoline additive in many countries.

Physicochemical and Environmental Considerations

Lead in gasoline was emitted typically at approximately 24,000 μg/m^3 at the tailpipe in the 1970s (Dzubay et al., 1979), and urban air contained lead at about 1-10 $\mu g/m^3$. A combination of air dilution and atmospheric fallout through dry and wet deposition accounts for the difference.

As described in detail elsewhere (EPA, 1986a), air lead from gasoline depends in a complex way on distances from vehicular traffic, lead content of gasoline, and mixing with the atmosphere. In closed spaces, such as garages and tunnels, air lead concentrations are well above those of open areas. Exhaust lead is discharged in such forms as halides and oxides, but these are eventually converted to the sulfate. Once the lead is dispersed, physical and chemical changes occur, including changes in particle size distributions, chemical changes from organic to inorganic lead, and chemical changes in the inorganic species themselves.

Most exhaust lead is deposited near its vehicular source (e.g., Reiter et al., 1977; Harrison and Laxen, 1981), whereas undeposited matter reaches a stable particle form within 100-200 km of its source. Particles approximately 10 μm in diameter are deposited over a broader distance, and there is long-range transport of particles less than 0.1 μm in diameter for up to about a month (e.g., Chamberlain et al., 1979).

General Characteristics of Exposure

The amount of lead consumed in the manufacture of antiknock additives for leaded gasoline has been enormous. In the United States, EPA (1986a) has estimated total consumption for 1975-1984 as 1.1 million metric tons. It has also estimated (EPA, 1986a) that 4-5 million metric tons have been deposited in the environment in the United States since introduction of alkyl lead additives in the mid-1920s.

Leaded-gasoline use is being phased out in the United States, as a result of a series of regulatory actions beginning in 1973, when EPA promulgated the requirement for unleaded gasoline for use in vehicles with catalytic converters (EPA, 1973), devices that would be damaged by lead. In 1982, EPA promulgated new rules (EPA, 1982) that switched the basis of the standard from an average lead content of all gasoline to an average lead content of leaded gasoline and set a limit of 1.1 g/gal in leaded gasoline. On August 2, 1984, EPA proposed to reduce the permissible amount of lead in gasoline to 0.1 g/gal, effective

January 1, 1986. The final regulation, issued in early 1985 (EPA, 1985), imposed an interim limit of 0.5 g/gal in June 1985. As a result of those actions, lead use in gasoline declined from 175,000 metric tons in 1976 to less than 4,000 metric tons in 1988.

The decline has been accompanied by a remarkable parallel decline in the mean blood lead concentration of the U.S. population (see Figure 1-3). Figure 3-8 shows the adjusted mean blood lead concentrations in the NHANES II survey—controlled for age, race, sex, income, degree of urbanization, region of the country, occupational exposure, dietary intake, and alcohol and tobacco consumption—plotted against national gasoline lead use.

The various analyses of the blood lead-gasoline lead relationship, via the NHANES II data set (Annest et al., 1983; Schwartz and Pitcher, 1989) and data from an isotopic-lead study (Facchetti and Geiss, 1982), show that gasoline lead, via both direct inhalation and exposure to fallout, can account for 50% or more of total blood lead concentration at the earlier air lead contents attributable to gasoline consumption.

Scope of the Problem

Although leaded gasoline is being phased out in the United States, huge quantities of deposited lead remain in environmental compartments from the many decades of use of leaded gasoline starting in the 1920s. Using linear and logistic regression analysis, Schwartz et al. (1985) have estimated the U.S. short- and long-term effect of leaded gasoline in terms of decreases in blood lead concentrations due to the phasing out of leaded gasoline, projected to 1992. The estimates are presented in Table 3-5 for children 0.5-13 years old.

Lead in Dust and Soil

This section deals with a major pathway of exposure to lead: lead in dust and soil. Lead in those media are now recognized to be derived from several sources, including leaded paint and atmospheric lead. The magnitude of the pathway and of the associated health hazards has been documented only recently.

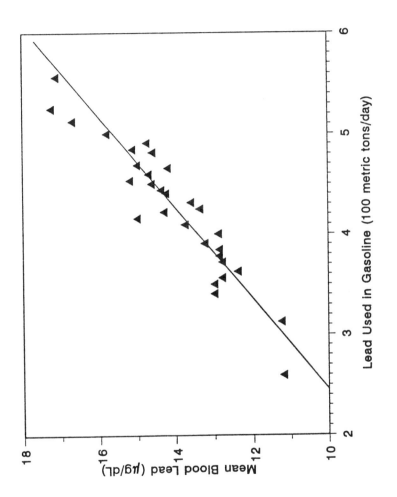

FIGURE 3-8 U.S. mean blood lead vs. lead used in gasoline. Source: Schwartz and Pitcher, 1989.

TABLE 3-5 Estimated Reductions in Blood Lead Content Because of Phaseout of Leaded Gasoline, Children 6 Months to 13 Years Old[a,b]

Year	No. Children with Blood Lead over 10 µg/dL	No. Children with Blood Lead over 15 µg/dL
1986	4,929,000	1,726,000
1987	4,595,000	1,597,000
1988	4,261,000	1,476,000
1989	3,918,000	1,353,000
1990	3,637,000	1,252,000
1991	3,283,000	1,125,000
1992	3,215,000	1,098,000

[a]Estimated by logistic regression analysis.
[b]Data from Schwartz et al., 1985.

Physicochemical and Environmental Considerations

Technically, dusts and soils are discrete physicochemical substances. However, both are stable, immobilizable, and relatively permanent depositories of contaminating lead (Yankel et al., 1977; Angle et al., 1984; Brunekreef, 1984; CDC, 1985; EPA, 1986a; ATSDR, 1988). Soils reflect precursor geology; dusts reflect atmospheric fallout and other deposition. Dusts have a wider range of particle sizes, including very small particles. Dusts and soils interact differently in human exposures and numerous recent studies have documented the different types of interactions. Some are discussed below.

Lead is present in dusts and soils at potentially toxic concentrations, primarily because of its use in leaded paint and its fallout from the air (Nriagu, 1978; Brunekreef, 1984; Duggan and Inskip, 1985; EPA, 1986a; ATSDR, 1988). It is often difficult to apportion lead content in soils and dusts accurately to either paint or atmospheric lead, but one or the other dominates in some circumstances and both contribute importantly in other circumstances.

Dusts and soils in remote communities near to primary lead smelters often are enriched in lead from atmospheric fallout due to these operations, either via direct emission or via re-entrainment of already highly contaminated media (e.g., Yankel et al., 1977; Angle et al., 1984; CDC, 1986a,b; EPA, 1986a; ATSDR, 1988). Soils and surfaces next to high-density roadways have similarly been documented as being heavily contaminated by vehicular exhaust emission of lead particles and other substances (Nriagu, 1978; Brunekreef, 1984; Duggan and Inskip, 1985; EPA, 1986a).

In inner-city areas with tracts of old, deteriorating housing, dusts and soils in adjacent interior and exterior sites are heavily contaminated with leaded paint (Sayre et al., 1974; Charney et al., 1983; Chisolm et al., 1985; Clark et al., 1985; Bornschein et al., 1987; EPA, 1986a; ATSDR, 1988). Studies of leaded-paint weathering in areas low in automobile density have identified contamination of soils and dusts and have shown a contamination pattern consistent with the presence of paint lead. Some inner-city neighborhoods also receive lead fallout from various mobile sources (vehicular exhaust) and stationary sources (secondary smelters, battery plants, and municipal incinerators).

A number of studies have attempted to measure the paint contribution to lead in dusts and soils; some are summarized in Table 3-6. Exterior-paint lead on homes, outbuildings, and such other outside entities as bridges will be transferred to adjacent soils and dusts (Ter Haar and Aronow, 1974; Linton et al., 1980; Landrigan et al., 1982; Fergusson and Schroeder, 1985; Schwar and Alexander, 1988). Movement of soil lead and interior-paint lead to interior dusts has also been documented (e.g., Clark et al., 1985; Bornschein et al., 1987, 1989). The removal of leaded paint from various surfaces warrants extreme caution.

Concentrations of lead in soils in rural areas of the United States are typically less than 30 μg/g of soil. In areas affected by lead mining, industrial emissions or vehicular traffic emitting leaded exhaust, such concentrations can increase by a factor of hundreds or even thousands. Automobile emissions account for most of the lead in soil and dust in suburban and rural areas (Nriagu, 1978; EPA, 1986a), whereas paint, atmospheric fallout from vehicular exhaust, and stationary sources account for most of the lead in urban dust and soil.

TABLE 3-6 Representative Studies of Contribution of Leaded Paint to Lead in Dusts and Soils

Study Site	Study Design	Results	References
Lead-painted frame and brick homes, Detroit, Mich., area	Soil lead vs. distance from test buildings (N=18 each type)	Lead in soil 2 ft away 5 times higher than in samples 10 ft away	Ter Haar and Aronow, 1974
Lead-painted rural barns and urban homes with leaded paint	Soil lead vs. distance from two painted building types	Similar soil content for both building types	Ter Haar and Aronow, 1974
Outside areas around homes in small town	Dust lead samples, curbside vs. at building line; electron microscopic chemical and surface analysis with element markers	25–85% of dwelling-line particles were paint flakes	Linton et al., 1980
House dust from homes, Christchurch, New Zealand	Housedust lead as function of home age and type: painted surface, brick, etc.	In homes with leaded paint in interiors, paint lead adds 45% to total dust lead content	Fergusson and Schroeder, 1985
Neighboring soils, bridge, Mystic, Conn.	Distance-stratified soil lead (1-cm layer) from bridge during and after lead removal	Soil lead 8,127 µg/g at bridge; 3,272 µg/g up to 30 m away; 457 µg/g 30–80 m away, and 197 µg/g 100 m away	Landrigan et al., 1982

Variable-quality housing, Cincinnati, Ohio	Dust lead (internal and external) and dust fall rate vs. house age, paint lead, and condition	All measures much higher in poor housing with paint lead	Clark et al., 1985
Variable-quality housing, Cincinnati, Ohio	Statistical analysis (structural equation modeling) of lead pathway in 18-month-olds	Paint lead and external-dust lead explain 52% of dust-lead variation; paint lead correlated with external-dust lead	Bornschein et al., 1987
Various residential areas and homes undergoing deleading	Analysis of housedust lead or child blood lead from paint dust generated during and after paint lead removal	Such dust formation has substantial effect on child exposure and blood lead	Rey-Alvarez and Menke-Hargrave, 1987; Amitai et al., 1987; Farfel and Chisolm, 1987; Rabinowitz et al., 1985a; Charney et al., 1983
Playground areas at schools undergoing lead-paint removal and repainting	279 schools in London, England, tested for play-area dust lead before, during, and after removal of old paint	Substantial increases in play-area dust lead after old-paint removal	Schwar and Alexander, 1988

Atmospheric lead from exhaust and stationary sources is transferred to soils through both dry (Friedlander, 1977; Schack et al., 1985) and wet (Lindberg and Harriss, 1981; Talbot and Andren, 1983; Barrie and Vet, 1984) depositional processes. The proportions of the type of contribution to the total deposition of lead vary markedly; wet depositional processes can account for up to 80% (Talbot and Andren, 1983).

Lead deposition onto soils from vehicular exhaust declines exponentially with distance from the roadway (EPA, 1986a). Much of this lead is immobilized in the top 5 cm of undisturbed soil (Reaves and Berrow, 1984) according to a complex function of geochemistry and pH (Olson and Skogerboe, 1975; Zimdahl and Skogerboe, 1977).

General Characteristics of Exposure

The extent to which lead in dusts and soils is translated into blood lead and later to some adverse effect depends first on the nature of the exposed population. Among sensitive populations, young children are most exposed to lead via dusts and soils, because they commonly put their hands in their mouths and often mouth or ingest contaminated soils.

The relationship of lead in dusts and soils to blood lead in young children has been the subject of various epidemiologic studies in urban areas and in rural areas that have stationary sources, such as smelters. The studies have been examined by EPA (1986a) and ATSDR (1988); the key studies include those of Yankel et al. (1977), Angle et al. (1984), CDC (1986a,b), Bornschein et al. (1987), and Rabinowitz and Bellinger (1988).

The complexities of and relation among sources and pathways have been quantitatively examined by Bornschein and co-workers (1987) on the basis of longitudinal environmental epidemiologic assessments of 18-month-old inner-city children. The children had substantial but not exclusive lead exposure through mobilizable leaded paint in poor-quality housing. The authors found that blood lead concentrations are influenced by the presence of lead in interior and exterior dust through hand pickup of lead in exploratory activity; dust lead is controlled by exterior dust (sampled as surface scrapings) and interior paint; and the lead concentration gradient works from the exterior to the interior, so the external contamination around a child's residence markedly affects the interior contamination and thus the exposure risk.

Other studies have shown a strong association of soil and interior dust lead with children's blood lead concentration; the various regression-slope estimates for earlier reports have been tabulated by EPA (1986a). The estimates cover a broad range, but suggest that detectable effects of these media on blood lead concentrations would occur at dust and soil concentrations of 500-1,000 μg/g (CDC, 1985). Particle size, chemical species of lead, and type of soil and dust matrices are important modifiers of the soil and dust lead hazard, because they influence lead intake and absorption (Roy, 1977; Barltrop and Meek, 1979; Heyworth et al., 1981; Healey, et al., 1982; Dornan, 1986; Koh and Babidge, 1986; Steele, et al., 1990). For example, particles of different lead-based paints are likely to have different solubilities (e.g., higher for the older lead carbonate paints, lower for the newer lead chromate paints), different particle sizes (large for paint chips ingested directly from walls or window sills, smaller for particles settled on dust or soil, very fine for particles formed by chalking or burned off walls by poorly applied heat guns), and thus different bioavailability for young children.

Scope of the Problem

A total of 20 million or more housing units were built before 1940; they are the units most likely to have old flaking, chalking, and weathering highly leaded paint that is being transferred to soils and becoming dust. The estimated numbers of young children discussed earlier as exposed to leaded paint are simultaneously exposed to lead in dust and soil. Added to the lead in soil and dust from paint in urban areas are the sizable amounts of lead from fallout from heavy vehicular traffic and stationary sources.

Schwartz et al. (1985) have used linear and logistic regression analysis of declines in child blood lead concentration associated with phasing out of leaded gasoline to estimate that 1.35 million children in 1989 and 1.25 million children in 1990 will have blood lead concentrations below 15 μg/dL as a result of the control action. Those numbers, when combined with projections to 1992, reflect a substantial change in dust lead concentrations associated with the decrease in fallout.

Lead in Drinking Water

Lead in tap water—consumed in the home, offices, other worksites, and public buildings—can be a particularly important source of lead exposure of young children, pregnant women, and other people (Moore et al., 1985; Levin, 1986, 1987; Ohanian, 1986; ATSDR, 1988). The potentially major role of tap-water lead in overall human exposure has long been recognized in Europe and older areas of the United States, but only recently has the full scope of the U.S. water-lead problem been examined. Reasons include the complex, heterogeneous nature of water supplies, the absence of detailed current survey data, and the relative exclusion of tap-water lead in environmental exposure assessment of lead-poisoning cases.

Physicochemical and Environmental Considerations

Tap-water lead concentrations are highly variable from house to house and tap to tap, because of differences in soldering, temperature, and water use. Any attempt to measure exposure or compliance with a target concentration must rest on an adequate sample size. For example, Schock et al. (1989) used data from four communities and found that 225-625 samples were required to produce a sample mean within 20% of the population mean (95% confidence limit), depending on the community. If no more than 10% of subjects should be exposed above a given value, the number of samples needed for accurate statistical inference would be even higher.

Lead theoretically can enter tap water at any of several points in the delivery system. A water-treatment plant distributes finished water with very little lead (e.g., Levin, 1986) and little more is added through the distribution lines, but contamination of domestic tap water occurs at five kinds of points in or near residential, public, or office core plumbing: lead connectors (i.e., goosenecks or pigtails), lead service line, lead-soldered joints in copper plumbing, lead-containing drinking fountains and water coolers, and lead-containing brass faucets and other fixtures. A host of chemical and physicochemical variables affect the extent to which any or all of those sites contribute to the water lead content. The important variables include the relative corrosivity of the water (i.e., acidity, alkalinity, and ion content), the standing time of water in contact with leaded surfaces, the age of lead-soldered joints and other leaded components, the quantity and surface area of lead sites, and the temperature of water in contact with lead surfaces.

In general, the problem with lead connectors and service pipe is associated principally with old housing, built around 1920 or before, in older northeastern American cities, particularly such New England cities as Boston (Worth et al., 1981; Karalekas et al., 1983). Since 1986, federal law has prohibited lead for these uses in new construction.

Solder-lead leaching varies with the age of plumbing and diminishes as the solder sites age, a process assumed to take about 5 years. The extent of lead leaching is strongly affected by the acidity of the water (i.e., pH), as seen in Table 3-7, which summarizes EPA data on pH and age of homes. With corrosive water (pH less than 6.4), it can be seen that soldered joints more than 5 years old still leach sizable amounts of lead in first-draw samples. Copper plumbing with lead-soldered joints came into widespread use in the United States and other developed countries in the 1950s.

Brass faucets and other fixtures containing alloyed lead at various percentages, even below current permissible percentage (8%), can contribute to tap-water contamination (Samuels and Meranger, 1984; Schock and Neff, 1988; Gardels and Sorg, 1989). Gardels and Sorg (1989) reported that newer brass faucets could contaminate standing water closest to the fixtures (less than 250 ml) at over 10 μg/L, an action concentration promulgated by EPA (1991).

In public facilities that serve young children and other sensitive populations, such as kindergartens and elementary schools, additional exposure to lead in tap water can occur (Levin, 1986; ATSDR, 1988; EPA, 1990c). Patterns of water use in schools potentially can allow greater exposures than in homes. For example, lead leaching is at its maximum into standing water, i.e., water generated overnight, during weekends, and during holiday and summer vacation periods. Water contamination in schools and the like can occur in water coolers and fountains, as well as in the expected core plumbing and fixtures (ATSDR, 1988; Gardels, 1989).

TABLE 3-7 Percentage of Variably Collected Water Samples Exceeding Lead at 20 µg/L at Different pH, by Age of House

		Samples with Lead over 20 µg/L	
Age of House, yr	pH	First Flush	% Fully Flushed (2 min)
0-2	<6.4	93	51
	7.0-7.4	83	5
	>8.0	72	0
More than 2, less than 6	<6.4	84	19
	7.0-7.4	28	7
	>8.0	18	4
6 or more	<6.4	51	4
	7.0-7.4	14	0
	>8.0	13	3

Source: ATSDR, 1988. Data from EPA, 1987.

General Characteristics of Exposure

Tap-water lead affects different groups in different ways. For example, lead-contaminated water can be used in infant formula and in beverages for older children and can be consumed directly. Tap-water lead can be ingested in foods cooked in lead-contaminated water. Furthermore, food surfaces can bind and concentrate water lead (Smart et al., 1981; Moore, 1985).

Lead in tap water is much more bioavailable than lead in food, because it is often consumed during semifasting (between meals) or after fasting (overnight) conditions. According to the data of Heard and Chamberlain (1982) and Rabinowitz et al. (1980), adults' fasting absorption rates can be 60% or higher, compared with rates of 10-15% in association with meals.

The marked reductions in blood lead concentrations of water-consumers in Glasgow and Ayr, Scotland, after water-treatment steps to reduce corrosivity (see, e.g., Moore et al., 1981; Sherlock et al., 1984; Moore, 1985) constitute convincing evidence of an impact of water lead on blood lead concentration. Figure 3-9 shows blood lead concentration distributions for two periods in a single group of mothers mointored before and after change in tap-water pH in Ayr, Scotland (Richards and Moore, 1984; Sherlock et al., 1984). Significant declines were observed in blood lead values as water was treated between 1980 and 1982.

With respect to case reports of lead intoxication associated with tap water, Cosgrove et al. (1989) reported lead intoxication—blood lead concentrations ranging up to 45 µg/dL—in a toddler found to have been exposed to lead solely from tap water, which entered the home through new copper plumbing with lead-soldered fittings. First-draw water samples averaged 390 µg/L and were as high as 1,080 µg/L.

Environmental epidemiologic studies have attempted to analyze the quantitative relation of tap-water lead to blood lead concentrations in both infants and adults. Some of the studies considered both dietary and water data; others examined only tap-water lead. As can be seen in Table 3-8 and in the very detailed Table 11-51 of EPA (1986a), for mainly first-draw water samples (except Sherlock et al., 1984), the relation of blood lead to tap-water lead is complex and a function of the concentration range of water and blood lead in the cohort. Over a broad range of water concentrations, well above 100 µg/L, the relation is curvilinear (e.g., Worth et al., 1981; U.K. Central Directorate on Environmental Pollution, 1982; Sherlock et al., 1984); it becomes linear at the low end of water content, i.e., less than 100 µg/L (Pocock et al., 1983). At the higher concentrations, the relation is best fitted through logarithmic or cube-root expressions.

Scope of the Problem

Table 3-9, based on data from Levin (1986) and ATSDR (1988), shows the numbers of children up to 13 years old who were at risk of exposure to lead from domestic plumbing. As indicated, 1.8 million children up to 13 years old lived in homes with newly installed lead-soldered plumbing (that is less than 2 years old), of whom 0.7 million

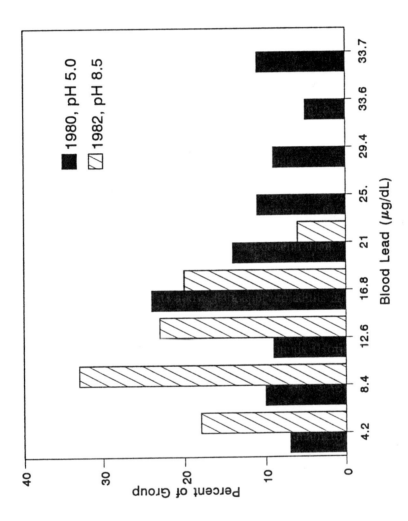

FIGURE 3-9 Blood lead in Ayr, Scotland, mothers before water treatment (1980, pH 5.0) and after water treatment (1982, pH 8.5). Source: Adapted from Richards and Moore, 1984.

TABLE 3-8 Selected Studies of Relation of Blood Lead to Tap-Water Lead

Study Details	Form of Model	Reference
524 Boston residents; water lead up to 1,108 μg/L; blood lead up to 71 μg/dL	ln(blood lead) = ln[0.041 (water lead) − 0.000219 (water lead)^2]	EPA (1986a) analysis of Worth et al., 1981
128 Glasgow mothers; water lead up to 1,060 μg/L; blood lead up to 39 μg/dL	Blood lead = 13.2 + 1.18 (water lead)$^{1/3}$	U.K. Central Direct., 1982
126 Glasgow infants of above mothers	Blood lead = 9.4 + 2.4 (water lead)$^{1/3}$	U.K. Central Direct., 1982
114 Ayr, Scotland, mothers before and after water treatment	Blood lead = 5.6 + 2.62 (water lead)$^{1/3}$	Sherlock et al., 1984
7,735 middle-aged British men; water lead less than 100 μg/L	Blood lead = 14.48 + 0.062 (water lead)	Pocock et al., 1983

Source: Adapted from EPA, 1986a, Vol. III.

TABLE 3-9 Estimated Numbers of Children at Risk of Exposure to Lead in Household Plumbing

Housing Type	Population at Risk
New Housing[a]	
8.8 million people in new housing with lead soldered piping:	
(8.8 million) (7.6% of population less than 5 years old)	0.7 million
(8.8 million) (12.8% of population 5-13 years old)	1.1 million
Total number of children at risk in new housing	1.8 million
Old Housing[b]	
If one-third of units built before 1939 contain lead pipes,[c] then (0.33) (0.29) = 10% of housing has lead pipes:	
(0.10) (17.8 million children less than 5 years old)	1.8 million
(0.10) (30.1 million children 5-13 years old)	3.0 million
Total number of children at risk in old housing	4.8 million

[a]Data from Levin, 1986; based on 9.6 million in new homes, of which 92% have metal plumbing.
[b]Data from U.S. Bureau of the Census, 1985.
[c]Data from David Moore, Office of Policy Development and Research, HUD, Submissions to ATSDR, January 1987, and EPA.
Source: ATSDR, 1988.

were less than 5 years old. The corresponding tally for old leaded plumbing is 4.8 million, of whom 1.8 million were under 5 years old. The two groups yield a total of 6.6 million children.

According to Levin's (1986) analyses, 42 million U.S. residents receive water from public supplies having lead concentrations above 20 μg/L. ATSDR (1988) estimates that 3.8 million of those are chil-

20 µg/L. ATSDR (1988) estimates that 3.8 million of those are children less than 6 years old.

Levin (1987) has used regression-analysis methods described by Schwartz et al. (1985) to estimate that 240,000 children less than 6 years old will have blood lead concentrations over 15 µg/dL, in part because of exposure to lead in tap water.

Lead in the Diet

Physicochemical and Environmental Considerations

Lead contaminates food through various pathways: deposition of airborne lead, binding of soil lead to root crops, use of lead-contaminated water and equipment in processing, use of lead-soldered cans for canned foods, and lead leaching from poorly made lead-glazed food and beverage containers.

Lead is readily deposited on leaf surfaces of edible plants (e.g., Schuck and Locke, 1970) and accumulates over the life of the crop. The deposition rate in areas with high air lead content can measurably increase the lead content of leafy crops, and such surface contamination is difficult to remove by either harvest washing or rainfall (Page et al., 1971; Arvik and Zimdahl, 1974).

Transfer of lead from soil to edible roots is a complex function of physicochemical factors that govern the plant uptake of lead, including those mentioned earlier in this chapter. Camerlynck and Kiekens (1982) reported that normal soils contain exchangeable lead at approximately 1 µg/g or less, and presumably some portion of the mobile lead will bind to plant roots.

Lead in processing water can sometimes be the major contributor to dietary lead (Moore et al., 1979; Smart et al., 1981). However, the more common source of food contamination in processing is the use of lead-soldered cans. When lead is used as seam solder, the material can spatter on the interior of the can or the toxicant can migrate to the canned-food matrix itself. Acidic foods induce more lead release from the soldering material, although the leaching phenomenon also occurs with relatively low-acidity foods, such as corn and beans, and in all cases the total amounts liberated are a function of the shelf-life of the canned goods. Lead release is accelerated by contact with oxygen once a can is opened. Lead in wine has been shown to be a potentially important source of dietary lead exposure (e.g., Elinder et al., 1988).

Pottery, dinnerware, and other ceramic items are used to store foods. If containers so used have been made with poorly fired leaded glazes, lead can migrate from them into the food (extensively discussed in Lead in Housewares, U.S. House, 1988; see also Wallace et al., 1985). Key factors affecting lead release include characteristics of the glaze, the temperature and duration of food storage, and the acidity of the food. Lead can also be released on extended scrubbing and cleaning of even well-prepared glazes.

Commercial American products have led to fewer problems in this regard than commercial products from other countries—such as countries in southern Europe and Latin America and mainland China—or items made by artisans and hobbyists. Most cases of lead toxicity have been associated with repeated use of vessels with problematic glazes or with prolonged food storage.

Glassware is often decorated with decals or decorative surfaces that contain lead. Those surfaces have a potential for exposure through contact with young children's lips and mouths.

Characteristics of General Exposure

The contribution of foods and beverages to body lead burden, as reflected in blood lead, has been measured in epidemiologic surveys of infants, toddlers, and older people (EPA, 1986a). Various studies have shown that dietary lead can contribute substantially to blood lead in complex ways that reflect the influence of the tap-water lead component, dietary habits, and individual differences in lead toxicokinetics (see, e.g., EPA, 1986a).

Ryu et al. (1983, 1985) found that dietary lead affects blood lead in infants in a simple linear fashion, at least in moderate exposures. Sherlock et al. (1982) and the U.K. Central Directorate on Environmental Pollution (1982) examined the relation in infants and mothers via a duplicate-diet survey. Blood lead in the infants in the U.K. study was related to dietary lead by both linear and cube-root functions, whereas Sherlock and co-workers found a cube-root relation for mothers and infants. The relation becomes curvilinear when intake exceeds 100 µg/day. The slopes of the curves (µg/dL of blood vs. µg/day), which can be estimated from the above studies for an intake of 100-200 µg/day, are 0.034 for adults (Sherlock et al., 1982), 0.06 for infants (Sherlock et al., 1982), 0.053-0.056 for infants (U.K. Central Directorate, 1982), and 0.16 for infants (Ryu et al., 1985). The relation of Ryu and co-workers has the steepest slope and is based on the lowest average lead intake; the slope might level at much higher lead intakes.

Dietary intakes of lead are being reduced in the United States (EPA, 1986a; ATSDR, 1988). For 2-year-olds, for example, there was a decline of approximately 75%, from 52.9 to 13.1 µg/day, from 1978 to 1985. It should be remembered that there is a distribution of lead content about the average and that the diet-survey numbers are based on relatively small samples, compared with the volume and diversity of the U.S. food supply. Several important factors in the decline include the domestic phaseout of lead-seamed beverage and food cans and the reduction in movement of lead to agronomic crops, as a result of a lowering exposure of growing crops to air lead. The latter is associated with the phaseout of leaded gasoline and tighter stationary-source regulations.

Table 3-10 shows U.S. production of lead-seamed cans in 1980 and 1988; these are production figures supplied by a trade group and do not reflect independent surveys of lead-seamed cans on grocery-store shelves. The latter would include some carryover from past years' production, depending on canned-food shelf life.

Scope of the Problem

As noted by Mushak and Crocetti (1989), virtually all sensitive populations are exposed to some lead in food, owing to the relatively centralized food production and distribution system in the United States and other developed nations. They also estimated on the basis of food-lead concentration distribution profiles, adjustments for lead reduction in foods, intakes from other lead sources—that approximately 5% of the 21 million U.S. children less than 6 years old, or 1 million children,

TABLE 3-10 Changes in Percentage of Lead-Soldered Food and Soft-Drink Cans (millions of cans)[a]

Year	Category	Total No. Cans (2-piece + 3-piece)	3-Piece Cans		No. Lead-Soldered, millions	% of Total
			No.	% of Total		
1980	Food and soft-drink cans	54,173	30,568	56.4	25,433	46.9
	Food cans	28,432	26,697	93.9	24,405	85.8
	Soft-drink cans	25,741	3,871	15.0	1,028	4.0
1988	Food and soft-drink cans	73,001	19,062	26.1	1,626	2.2
	Food cans	28,071	19,062	67.9	1,626	5.8
	Soft-drink cans	44,930	0	0	0	0

[a]Data from Can Manufacturers Institute, unpublished material.

will have increases in blood lead concentrations because of intake of lead in food.

Although the main theme of this report is ordinary environmental exposure to lead, occupational settings present the highest, continuous lead exposures of all. Furthermore, workers transport lead from work to their homes, where their families, including children, are exposed to it (e.g., Baker et al., 1977; Milar and Mushak, 1982).

Various traditional customs and medications can result in high lead exposures. Reports of such exposures are numerous in the clinical literature from various regions—Arab countries (Aslam et al., 1979), the Indian-Pakistani subcontinent (Pontifex and Garg, 1985), China (CDC, 1983a), and Latin America (Bose et al., 1983; CDC, 1983b; Trotter, 1985; Baer et al., 1987). The preparations often include lead compounds as major or principal ingredients, so the poisoning potential is high.

In the United States, the most familiar type of lead-containing preparation is a Mexican-American folk preparation that contains lead oxides (Bose et al., 1983; Trotter, 1985; Baer et al., 1987). Greta (lead (II) oxide) and azarcon (mixed-valence lead tetroxide, $PbO_2 \cdot 2PbO$) are used to treat digestive disorders; their use produces diarrhea or vomiting. Use of these medicines is widespread and can result in serious lead poisoning in children.

SUMMARY

It is difficult to rank sources of lead exposure by their importance for health by such simple criteria as numbers of affected persons. Simultaneous exposure to multiple lead sources is inevitable; different sources of lead are often associated with different degrees of lead poisoning, which would make it necessary to rank by effect severity, as well as frequency; and sources differ in distribution among sensitive populations. An alternative is to provide a ranking by relative overall impact, which includes the potential of a source for the most severe poisoning, its relative pervasiveness, estimates of numbers of persons exposed to it, and the relative difficulty of abating it.

The sensitive populations within the general, nonoccupational sector are preschool children, fetuses (via maternal exposure), and pregnant women (as surrogates for fetuses). On the basis of overall public-health impact on those populations, sources can be combined into two groups. Lead in paint, lead in dusts and soils, and lead in drinking water constitute the more important group today. In that group, leaded paint ranks first in importance for young children, followed closely by lead in dusts and soils, and then by tap-water lead. For adults, tap-water lead is probably the exogenous source of most concern. (Endogenous exposure to lead can occur when subjects mobilize lead and calcium from bone; this typically occurs in adults or in children who break bones.) In the United States, leaded gasoline at present concentrations and dietary lead make up the second group, of somewhat less concern. These statements of importance are relative; they do not imply that any specific source is unimportant as a contributor to lead body burdens or to earlier effects in populations as a whole. The body combines lead absorbed from all sources into one dose.

The phasedown of leaded gasoline is greatly reducing the input of lead to environmental compartments. However, the inventory of 4-5 million metric tons of lead still in the environment because of past leaded-gasoline use will continue to contribute to the risk of exposure of sensitive populations. Outside the United States, various approaches to leaded-gasoline control are being taken, from modest control actions to phaseout and phasedown regulations.

Leaded paint (and its transport to dusts and soils) is a major national source of exposure of children. Dust and soil lead comes from leaded-paint transfer and atmospheric fallout, and many studies have documented its contribution to lead body burdens of young children. Quantitative assessments of the relative contributions of dust and soil lead to total body lead, such as blood lead concentration, have been the subject of diverse studies. In addition, particle size, chemical species of lead, and soil and dust matrices are important modifiers of the soil and dust lead hazard eventually reflected in lead intake and absorption.

Pathways of exposure to tap-water lead are multiple: direct drinking, beverages prepared with contaminated water, and foods cooked in lead-contaminated water. Patterns of leaded-water use can amplify toxicity risk. Ingestion on an empty stomach, a common occurrence, greatly increases the rate of lead absorption. The use of water in elementary schools and other child facilities is intermittent, with extended standing time over weekends and in vacation periods. That allows buildup of lead in fountains and water lines.

Most developed countries, including the United States, have complex food production and food distribution systems that permit lead contamination. Virtually everyone has some exposure to dietary lead, and lead concentrations in food can be quite high. But lead in foods of older children and infants has been reduced through phasing out of lead-soldered cans for milk and fruit juices and reduced input into food crops.

There does appear to be a persisting problem with lead leaching mainly from poorly made and lead-glazed food and beverage pottery. It could also be that even well-made vessels with lead glazes will lose lead through extended surface abrasion, as in scrubbing, washing, and rinsing.

4

Biologic Markers of Lead Toxicity

In the last few years, considerable interest has developed in discovering and validating new biologic markers for many toxic substances. Identifying new biologic markers has helped scientists to understand much better the mechanisms of toxicity. This has also been the focus for biologic studies of the mechanisms of lead toxicity. Biologic markers are indicators of events in biologic systems or samples. The National Research Council (NRC, 1989a,b) has classified biologic markers into three types—markers of exposure, of effect, and of susceptibility. A biologic marker of exposure is an exogenous substance or its metabolite or the product of an interaction between a xenobiotic agent and some target molecule or cell. A biologic marker of effect is a measurable biochemical, physiologic, or other alteration within an organism that, depending on magnitude, can be recognized as an established or potential health impairment or disease. A biologic marker of susceptibility is an indicator of an inherent or acquired limitation of an organism's ability to respond to the challenge of exposure to a specific xenobiotic substance. This chapter describes biologic markers of exposure, effect, and susceptibility for lead. It also establishes the biologic basis for the assessment of analytic techniques to monitor lead in sensitive populations, which will be described in Chapter 5.

BIOLOGIC MARKERS OF EXPOSURE

Any assessment of the toxicity associated with exposure to lead begins with measurement of the exposure. In practice, one assesses lead exposure through environmental or biologic monitoring techniques to examine markers of exposure. Lead exposure is the amount of lead (from whatever source) that is presented to an organism; dose is the amount that is absorbed by the organism (NRC, 1990). Various factors—such as blood flow, capillary permeability, transport into an organ or tissue, and number of active binding and receptor sites—determine the path of lead through the body and can influence the biologically effective dose. For example, lead inhaled in dust could be retained in the lungs, removed from the lungs by protective mechanisms and ingested, stored in bone, or eliminated from the body via the kidneys. Toxicity can be observed in the kidneys, blood, nervous system, or other organs and tissues. At any step after exposure, biologic markers of exposure to lead can be detected.

A key component of biologic monitoring of lead exposure is the toxicokinetic and physiologic framework that underlies such monitoring. Screening in the absence of knowledge about lead's in vivo behavior limits the interpretation of monitoring data for public-health risk. For example, if clinical management or regulatory actions are to be effective, the timing of lead exposure that is reflected in a typical blood lead value should be known, as should dose-response relations that link the body lead concentration with adverse health effects.

Lead Absorption

Humans absorb lead predominantly through the gastrointestinal and respiratory tracts. Little uptake occurs through skin, especially in nonoccupational exposures. Lead deposition and absorption rates in the human respiratory tract are complex functions of chemical and physical forms of the element and of anatomic, respiratory, and metabolic characteristics.

Inhaled lead is deposited in the upper and lower reaches of the respiratory tract. Deposition in the upper portion leads to ciliary clearance of lead, swallowing, and absorption from the intestine. Smaller lead particles, especially those less than 1 μm in statistically averaged diameter, penetrate the lower, pulmonary portion of the respiratory tract and undergo absorption from it.

Human studies (Chamberlain, 1983; EPA, 1986a) have shown that about 30-50% of inhaled lead is retained by the lungs (the range reflects mainly particle size and individual breathing rate). These studies have used unlabeled lead aerosol (Kehoe, 1961a,b,c), radiolabeled oxide aerosol (Chamberlain et al., 1978), lead fumes inhaled by volunteers (Nozaki, 1966), ambient air lead around motorways and encountered by the general population (Chamberlain et al., 1978; Chamberlain, 1983), lead salt aerosols inhaled by volunteers (Morrow et al., 1980), and lead in forms encountered in lead operations, fumes, dusts, etc. (Mehani, 1966). Most (over 95%) of whatever lead is deposited in the human pulmonary compartment is absorbed (Rabinowitz et al., 1977; Chamberlain et al., 1978; Morrow et al., 1980). Thus, the overall rate of uptake is governed by lung retention (i.e., 30-50%). Uptake occurs rapidly, generally in a matter of hours.

Evidence of complete and rapid uptake can be gleaned from analysis of autopsy lung tissue (Barry, 1975; Gross et al., 1975). The chemical form of inhaled lead appears to have little effect on uptake rate (Chamberlain et al., 1978; Morrow et al., 1980). Similarly, uptake is little affected by air lead concentration, even when it is greatly in excess of that commonly encountered in nonoccupational settings—up to 450 µg/day (Chamberlain et al., 1978).

The above data apply to adults and are relevant for the sensitive adult population, i.e., pregnant women. In the case of children, no studies have experimentally documented rates of direct uptake of lead from lungs. On anatomic grounds (Hofmann et al., 1979; Hofmann, 1982; Phalen et al., 1985) and metabolic grounds (Barltrop, 1972; James, 1978), however, uptake in adults should be greater than uptake in children.

In nonoccupationally exposed populations, lead uptake from the gastrointestinal tract is its main route of absorption. For adults, the lead content of foods, tap water, and other beverages is of main concern for lead exposure. For infants, toddlers, and older children, ingestion of lead-contaminated nonfood materials—e.g., dust, soil, and leaded-paint chips—is of additional concern. In some cases, such exposure can exceed that occurring through the diet (NRC, 1976, 1980; EPA, 1986a; WHO, 1987).

Various studies of gastrointestinal absorption of lead in adults as derived from measures of metabolic balance (Kehoe, 1961a,b,c) and isotope distribution (Hursh and Suomela, 1968; Harrison et al., 1969; Rabinowitz et al., 1974, 1980; Chamberlain et al., 1978) have documented that 10-15% of dietary lead is absorbed. The rate rises considerably, to as high as 63%, under fasting conditions (Chamberlain et al., 1978; Rabinowitz et al., 1980; Heard and Chamberlain, 1982). That suggests that lead in tap water and other beverages, which are often imbibed on an empty stomach, undergo higher uptake and pose proportionately greater exposure. Rate of lead uptake over the range of exposures likely to be encountered by the general population, up to at least 400 µg/day, seems similar (Flanagan et al., 1982; Heard et al., 1983).

Studies of lead bioavailability in the human intestine (Chamberlain et al., 1978; Rabinowitz et al., 1980; Heard and Chamberlain, 1982) have indicated that common dietary forms of lead are absorbed to about the same extent. Lead sulfide in one study was absorbed to the same extent as other forms, and in another study was absorbed to the same extent with meals, but during fasting was absorbed less than the chloride. Particle size difference might account for the absorption difference between fasting and meals.

Dietary lead absorption is considerably higher in children than in adults. Results of studies of both Ziegler et al. (1978) with infants and Alexander and colleagues (1973) with children indicate an absorption rate up to 50% from the intestinal tract.

Young children ingest nonfood lead through normal mouthing behavior and particularly through the abnormal, excessive behavior called pica. Substantial uptake of lead and systemic exposure occur because of the high concentrations of lead in such media as dust, soil and leaded paint (Duggan and Inskip, 1985; EPA, 1986a; WHO, 1987); the ingestion of perhaps 100 mg, or even more, of such media (Binder et al., 1986; Clausing et al., 1987; Calabrese et al., 1989; Davis et al., 1990); and a bioavailability of up to 30% (Day et al., 1979; Duggan and Inskip, 1985; EPA, 1986a). The higher intestinal absorption of lead seen in developing versus adult humans is commonly observed in other mammalian systems, including nonhuman primates (Pounds et al., 1978) and rodents (e.g., Kostial et al., 1978).

Percutaneous absorption of inorganic lead in nonoccupational populations is low. Moore et al. (1980) applied ^{203}Pb-labeled lead acetate to intact skin of adult volunteers and obtained an average absorption rate of only 0.06%. Lilley et al. (1988) applied lead as the metallic powder or nitrate salt solution to one subject's skin; it failed to increase the lead content of either whole blood or urine, but the lead content of sweat far from the area of application increased.

Lead has long been known to cross the placental barrier in humans and other species and become lodged in fetal tissues (Barltrop, 1969; Chaube et al., 1972; Buchet et al., 1978; Alexander and Delves, 1981; Rabinowitz and Needleman, 1982; Borella et al., 1986; Mayer-Popken et al., 1986). The question of when lead begins to enter the fetus during maternal exposure is important, but has not been fully answered. Data of Barltrop (1969) and Mayer-Popken et al. (1986) suggest that lead entry occurs by the third or fourth month; data of Borella et al. (1986) and Chaube et al. (1972) suggest that uptake occurs later.

Lead Distribution

Absorbed lead enters plasma and undergoes rapid removal to various body compartments: erythrocytes, soft tissue, and mineralizing tissue. Removal occurs over a matter of minutes (Chamberlain et al., 1978; DeSilva, 1981). If exposure is constant, a steady state eventually occurs.

Under steady-state conditions (i.e., stable exposure), plasma lead and erythrocyte lead are in equilibrium. The equilibrium fraction of lead in plasma is less than 1% and varies very little (Cavalleri et al., 1978; Everson and Patterson, 1980; DeSilva, 1981; Manton and Cook, 1984), but rises at blood lead concentrations of about 50 μg/dL or higher (DeSilva, 1981; Manton and Cook, 1984).

Lead is removed from whole blood, under steady-state conditions, with a half-life that depends on such factors as total body lead burden, age, magnitude of external exposure, and the method of measuring half-life (according to total circulating lead or absorbed exogenous fraction as measured by isotopic tracer).

Whole-blood lead measured with various protocols of experimental exposure has been found to have a half-life of about 25 days (Griffin et al., 1975; Rabinowitz et al., 1976; Chamberlain et al., 1978). That refers to the first, or short-term, component of blood lead decay. Actual measurements of half-life, commonly obtained through blood lead changes that occur with reduction in chronic exposure, yield various and generally much higher values than those obtained experimentally; actual measurements reflect a larger contribution of a long-term component, described as many months in half-life.

Early studies by Barry (1975, 1981) and Gross et al. (1975) showed that lead in most soft tissues is usually below 0.5 parts per million (ppm); age-dependent accumulation in kidneys (Indraprasit et al., 1974) and aorta (Barry, 1975; Gross et al., 1975) in nonoccupational populations has been reported.

Available data are not sufficient to show whether soft tissue concentrations have been declining in response to lower air and dietary lead uptake in recent years. Such changes would be registered more readily in the youngest segments of the population, where cumulative body burdens are smaller.

Studies that showed no lead accumulation with age in many soft tissues have been cross-sectional and theoretically would disguise the moderate accumulation that can occur with age but be offset by declining lead exposure in recent years.

The human brain, the principal target organ of lead exposure, has low concentrations of lead—less than 0.2 ppm (wet weight)—on a whole-organ basis when there has been no occupational exposure. Lead content can rise by a factor of several in people with high lead exposure (Barry, 1975). In subjects with lethal poisoning, whole-brain concentrations are above 1 ppm (Okazaki et al., 1963; Klein et al., 1970). Region-specific distribution of lead in the brain has been documented. The highest concentrations are in the hippocampus and frontal cortex (Okazaki et al., 1963; Niklowitz and Mandybur, 1975; Grandjean, 1978).

Barry (1975, 1981) showed that tissue lead concentrations were lower in infants than in older children. Those in older children were not materially different from those in adult women.

A large body of laboratory and clinical evidence shows that lead accumulates with age in human mineralizing tissue, i.e., bones and teeth. Accumulation appears to begin at birth (or even in utero) and continues until the age of 50-60 years, when it starts to decrease through some combination of dietary, metabolic, and hormonal changes (CDC, 1985; EPA, 1986a; Drasch et al., 1987; Drasch and Ott, 1988;

Wittmers et al., 1988). Total lead content in bone can reach 200 mg in nonoccupationally exposed adults and much higher in those occupationally exposed to large concentrations.

Drasch and Ott (1988) have confirmed that bone lead is cumulative at least from birth. Autopsy samples from infants less than 1 year old had a bone lead concentration half that of preschool children (0.33 vs. 0.62 ppm wet weight) and one-fifth that of people 10-20 years old (1.76 ppm). All bone types—cortical bone, such as midfemur, and trabecular bone, such as temporal bone and pelvis—were shown to accumulate lead, but the denser cortical bone had markedly higher concentrations in the two older groups. A sex-based difference in bone lead accumulation was observed in the oldest group for trabecular bone, males having statistically higher concentrations ($p < 0.05$). Recent measurements of bone lead in adult autopsy samples also documented continued accumulation in adulthood up to at least the age of 50 (Drasch et al., 1987; Wittmers et al., 1988). In the work of Drasch et al. (1987), temporal bone showed age-dependent accumulation throughout adulthood, including the 70s, whereas midfemoral and pelvic samples showed a plateau in middle age and then a decline. The latter decline was pronounced in females and was attributed to osteoporotic changes. Those data support the finding in analysis of NHANES II results that menopausal women have higher blood lead than younger women (Silbergeld et al., 1988). Men were estimated to have a significantly higher total skeletal lead burden than women—mean, 41.4 mg versus 24.1 mg. Comparison of recent analyses with data gathered 10 years earlier in the same laboratory and with identical methods indicated a marked decline of lead in femoral and pelvic samples across adult age groups, amounting to 30-50%.

In similar investigations, Wittmers et al. (1988) examined tibia, skull, rib, ilium, and vertebra from 134 hospital autopsies for lead content as a function of age, lateral and cross-sectional analytic symmetry, and bone composition. Lead content was symmetric in positional location, but not bone type. Lead concentrations rose with age in all sample types, and there was some longitudinal variation within a bone specimen, but not enough to preclude use of single measurements in bone analysis.

Lead Retention and Excretion

EPA (1986a) and ATSDR (1988) analyzed the retention and excretion of lead in humans and animals. Ingested lead that is not absorbed is lost through urinary and fecal excretion. Absorbed lead that is not sequestered in bone or some soft tissues is eventually eliminated through the kidneys or through biliary clearance into the intestine. Deposition in keratinizing tissues (nails and hair) is a minor elimination pathway.

On the basis of various experimental measurements (Kehoe, 1961a,b,c; Rabinowitz et al., 1976; Chamberlain et al., 1978), the following can be said:

- Urinary loss of lead in adults makes up about two-thirds of total elimination. Fecal lead loss (of lead arising from biliary elimination—i.e., endogenous fecal lead) makes up about one-third. About 8% of the total is eliminated through Hair and nails.
- Whole-body lead elimination over the short term removes about 50-60% of the newly absorbed lead, with a half-life in adult volunteers of about 20 days (Rabinowitz et al., 1976; Chamberlain et al., 1978). Of the deposited fraction, 50% (i.e., 25% of lead initially absorbed) is eventually eliminated.
- Infants and children retain 50% of ingested lead (Alexander et al., 1973; Ziegler et al., 1978).
- Infants (and perhaps preschool children) have slower elimination than adults (Thompson, 1971; Alexander et al., 1973; Chamberlain et al., 1978; Ziegler et al., 1978; EPA, 1986a).
- Lead elimination through urine might depend on concentration, as estimated by Chamberlain (1983) on the basis of results of studies that reported blood and urine values in adults.
- Whole-body lead retention in humans subjected to constant exposure is accounted for largely by skeletal accumulation.

Interactions of Lead with Nutrients

The toxicokinetics of lead in humans are affected by the metabolic and nutritional status of the exposed subjects. Nutrition and nutritional deficiencies are of prime concern in very young children, in whom increased lead exposure is concurrent with deficiencies in many interactive elements, especially calcium and iron (see Markers of Susceptibility). Interactive relations for lead have been reviewed elsewhere (Mahaffey and Michaelson, 1980; EPA, 1986a). Various child and infant nutritional-status surveys have documented iron deficiency in children under 2 years old, especially those in low socioeconomic classifications. They are also the children with the highest prevalences of high body lead (Yip et al., 1981; Mahaffey et al., 1982a; Mahaffey and Annest, 1986).

Similarly, various reports have shown a strong negative correlation between calcium intake and blood lead in children (e.g., Sorrell et al., 1977; Ziegler et al., 1978; Johnson and Tenuta, 1979) and adults (Heard and Chamberlain, 1982). In the analyses of Ziegler et al. (1978), the inverse association of blood lead concentration and calcium intake in infants was seen to extend into the low part of the range of adequate intake.

Other nutrients that interact inversely with lead exposure are zinc (Chisolm, 1981; Markowitz and Rosen, 1981) and phosphorus (Heard and Chamberlain, 1982).

Mathematical Models

Over the years, a number of attempts have been put forth to provide a quantitative, mathematical model of the relation of lead in exposure media to total and toxicologically active lead in the body, the in vivo compartmentalization of lead in the human body, the relation of lead in target tissues and organs to likely biologic markers of exposure and toxicity, and even the relation of direct dose biologic markers to markers of early effect. Modeling approaches to metals in general are described by Clarkson et al. (1988), and specific reviews of lead biokinetic modeling are provided by Mushak (1989) and EPA (1986a).

Models of in vivo toxicokinetics of lead differ greatly, both in their use of empirical data and in the types of lead exposure to which they are applicable. One can broadly group toxicokinetic models of lead into linear and nonlinear forms. We are interested here primarily in models that are applicable to low-concentration lead exposure.

Linear Models

Rabinowitz and co-workers (1976, 1977) used stable isotopic-lead distribution analyses in adult volunteers to develop a three-compartment model of lead disposition. The kinetically discernible compartments were blood (the most mobile, containing 2 mg of lead), soft tissue (of intermediate mobility, containing 0.6 mg of lead), and bone (the largest, most stable, with a half-life of a decade or more and sequestering most total body lead).

The modeling efforts of Kneip and co-workers (Kneip et al., 1983; Harley and Kneip, 1984) expanded earlier approaches to a multiorgan model that can be used to estimate deposition in children of different ages. Figure 4-1 shows the major components of the Harley and Kneip (1984) approach that depicts six tissue compartments.

Table 4-1 presents age-dependent estimates of lead half-lives for bone, kidney, and liver reported by Harley and Kneip (1984) for the age range of 1-20 years. Bone lead in children 1-6 years of age has a half-life that is only one-third that in older children and only about half that in 8-year-olds. In contrast (and as expected), soft-tissue lead half-lives are independent of age. Age dependence (over ages 1-15 years) of tissue burdens of lead was also estimated (see Table 4-2). Blood lead peaks at 2 years and then declines gradually. Bone lead estimates show that lead concentration in 1-year-old infants is about 60% of that in 7-year-olds—not greatly at odds with the laboratory ratio of 1:2 based on autopsy samples for about the same age interval.

The models referred to are for essentially steady-state exposure with associated complete mixing of the linked lead pools, and they use first-order kinetics. They are chronic-exposure modeling approaches and are not to be considered valid for acute lead poisoning.

Nonlinear Models

At low to moderate lead exposure, linear models of lead in humans appear to be as good as any other form of mathematical depiction. However, any model intended to be broadly applicable to higher exposures must account for the known empirical curvilinearity of blood lead as a function of some external lead concentration (e.g., in water or air) and multiple subpools of lead (e.g., in blood and in bone).

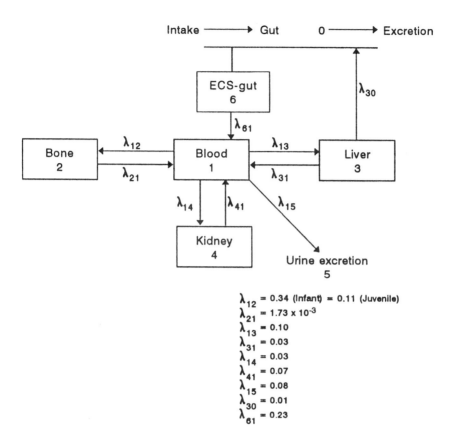

FIGURE 4-1 Linear toxicokinetic model of Harley and Kneip (1984). Model has six components, including initial extracellular space-gut compartment. Coefficients (λ) of compartment entry and exit are as indicated. Source: Kneip et al., 1983. Reprinted with permission from *NeuroToxicology*; copyright 1983, Intox Press.

The nonlinear model proposed by Marcus (1985) is an attempt to accommodate data that show that plasma lead manifests a concentration-dependent equilibrium with erythrocytes in humans and that blood lead concentration is nonlinear over a broad range of exposure. Workplace exposures represent the high end of the predictive range. The model provides a rather good fit for data from studies by DeSilva (1981) and Manton and Cook (1984) of subjects exposed over a broad range.

TABLE 4-1 Lead Half-Life Estimates by Age in Humans, Based on Linear Modeling

Age, yr	Tissue Half-Life, days		
	Bone	Kidney	Liver
1	1,135	10	23
3	1,135	10	23
6	1,135	10	23
8	2,560	10	23
15	3,421	10	23
20	3,421	10	23

Source: Adapted from Harley and Kneip, 1984, Table 4.

Chamberlain (1985) used a variant of nonlinear exposure models to rationalize the nonlinear relations of blood lead to media lead for a variable excretory function, dose-dependent urinary excretion, and thus incorporated dose-dependent transfer coefficients.

Biologic Monitoring

In recent years, the amount of lead entering the environment has declined because of regulatory or other risk-reduction actions and the body lead burden considered to pose an unacceptable risk of toxicity has declined. Those simultaneous decreases have shifted attention to small lead exposures and their associated subtle adverse effects. Small exposures add considerably to the complexity of interpreting lead pharmacokinetics. For example, the strong propensity for lead to accumulate in skeletal tissue and then be resorbed into blood with age increases the relative impact of endogenous input to blood at low concentrations. As noted in a recent report to Congress on childhood lead poisoning (ATSDR, 1988), high-dose exposure is not necessary for toxicity.

Despite increasing interest in low-concentration lead exposure, young children still sustain acute and subacute overt poisoning in areas of

TABLE 4-2 Estimates of Age-Dependent Lead Concentrations in Tissue, Based on Linear Modeling[a]

Age, yr	Lead Concentration		
	Blood, µg/dL	Bone, µg/g ash	Kidney, µg/g (wet weight)
1	11.9	35.5	0.7
2	16.2	38.1	1.0
5	14.5	51.0	0.9
7	13.0	57.9	0.8
10	10.4	57.6	0.9
15	11.3	41.7	0.7

[a]Based on 40 µg of lead absorbed (males).
Source: Adapted from Harley and Kneip, 1984, Table 12.

heavy lead contamination. It is still necessary, therefore, to consider the pharmacokinetics of lead at higher exposures.

Measurement of blood lead right after acute or subacute lead exposure is probably the only means of unambiguously establishing such exposure (Chisolm and Harrison, 1956; Chisolm, 1965; NRC, 1972, 1980; EPA, 1986a). That is because lead is rapidly absorbed into plasma and distributed to erythrocytes and vulnerable tissue sites while there is still a lag in response by most early-effect indicators, such as erythrocyte protoporphyrin.

Carton et al. (1987) showed the relative reliability and suitability of blood lead measures in an outbreak of acute lead poisoning caused by consumption of lead-contaminated flour in a Spanish village. A group of 136 poisoning patients whose blood lead concentrations were known at the outset were examined longitudinally. Of various exposure measures—erythrocyte protoporphyrin, coproporphyrin, urinary delta-aminolevulinic acid (ALA), blood delta-ALA-D, urinary lead, and blood lead—blood lead most closely corresponds to the severity of acute or subacute poisoning and to the overall laboratory and clinical pictures in the most severely affected people.

Whole Blood

For epidemiologic and clinical acceptability and utility, lead in the whole blood of chronically exposed populations remains the biologic marker of choice. It has been the traditional view that blood lead generally reflects fairly recent exposure, i.e., exposure 20-30 days before measurement. In cases of relatively stable exposure, however, such a short-term index of lead uptake is still of considerable utility.

The collection of blood lead in children in well-designed studies (either cross-sectional or longitudinal) is subject to problems of interpretation. Child blood lead is highly responsive to changes in exposure in the preceding 1 or 2 weeks. Most studies involve a recruitment phase that precedes blood lead collection (or the first blood collection in a longitudinal study) by a few days to a few months. The recruitment activity itself is an interaction with the child's family or caretakers that might alter their behavior and increase their awareness of potential lead exposure hazards. The primary exposure vector for young children is household dust and surface soil (even if the source is deteriorating lead-based paint), so changes in caretaker behavior that reduce dust exposure might cause a reduction in a child's blood lead concentration between the time of recruitment and the time of blood collection. Such changes include more frequent dusting and handwashing and more effective control of child access to dusty or dirty places. Changes in caretaker behavior are even more likely in longitudinal studies with repeated contacts between investigators and subjects or in cross-sectional studies in communities that have already had long-standing media coverage or community controversy about removal of leaded paint or lead-contaminated soil.

Blood lead concentration as a short-term measure of exposure is less accurate with subjects whose skeletal lead contributes substantially to total blood lead concentration. Estimating total body lead burden is complex and requires that all sources of lead be considered. Consequently, it might be expected that, in people (children and adults) with a high body lead burden lodged in bone, more of the lead in bone would contribute to blood lead concentration and be reflected in the long half-life of removal from blood. With regard to fractional contributions of recent versus cumulative lead exposure to blood lead, various and numerous studies have shown that the major component of a given total blood lead concentration in a young child or adult is recent input, and the influence of cumulative input increases as a function of age and exposure history (Duggan, 1983; Christoffersson et al., 1984; Harley and Kneip, 1984; Schwartz et al., 1985; Schütz et al., 1987a; Skerfving, 1988). One exception would be retired or reassigned lead workers who received heavy lead exposure in their working careers.

Table 4-3 summarizes data related to the toxicokinetic aspects of blood lead in children. Corresponding data on diverse adult populations are set forth in Table 4-4. The tables should be read for their implications for biologic monitoring, especially for low-dose exposures.

A number of conclusions are to be drawn from Table 4-3, including:

• Blood lead appears to stabilize in older children, at least enough to preserve rank order, especially when exposure is reduced.
• Rate of change of blood lead of infants and perhaps older children in response to changes in exposure appears to be a function of current exposure and accumulated body burden.
• Older children appear to preserve an earlier exposure history (in bone stores), as shown in rank ordering; there might be at least two lead compartments that contribute to blood lead concentration, one of which is large enough to preserve statistical association (consisting of lead in bone), although continuing exposure cannot be ruled out when blood lead concentrations are large.

Data in Table 4-4 can be summarized as follows:

• Men and women usually exposed to lead in air and diet have a lower rate of change in blood lead in response to exposure changes than those with little or no exposure.
• Bone lead can be an important source of steady-state blood lead concentration, even under conditions of ordinary exposure. Similarly, occupationally exposed people can accumulate a skeletal burden large enough to become the dominant source of blood lead concentration even after active exposure ceases.
• Blood lead clearance in substantially exposed people can be described by two components, one of which is rapid (1-2 months) and reflects soft-tissue lead, and one a longer-term bone component that is reflected in reduction of bone lead stores.

TABLE 4-3 Studies of Kinetic Behavior of Lead in Blood of Children

Study Group and Exposure	Half-Life, days	Comments (References)
Infants, middle class; ambient exposure	-	Blood lead very unstable for first 20 mo Rabinowitz et al., 1984)
Infants, middle class; ambient low exposure	5.6	Reanalysis of Ziegler et al. (1978) data; mean-time 8 days (half-life, 5.6 days) (Duggan, 1983)
Infants, low socioeconomic status; heavy ambient exposure	ca. 300	Reflects high body burden plus in utero uptake in urban setting (Succop et al., 1987)
Low-socioeconomic-status children of battery workers; secondary exposure	-	Rank order of group preserved over 5 yr; $r = 0.74$ (Schroeder et al., 1985)
General U.S. child population; varied exposure	-	Regression analyses of NHANES blood lead data showed 30-day (best-fit) lag with lead source (Schwartz et al., 1985; Annest and Mahaffey, 1984)
School-age English children; low exposure	-	Two blood lead sets, 20 mo apart; rank order preserved (Lansdown et al., 1986)
U.S. children, 4-12 yr old; increased ambient exposure	-	Rank order of serial blood lead measures generally preserved (David et al., 1982)

- Generally, blood lead half-life is highly variable because of such factors as age, metabolic variability, total body burden, and concentration and duration of exposure.

With careful attention to methodologic details, blood lead concentration analyses by competent laboratories can be used in general population surveys of trends in lead exposure. That has proved especially useful in relating recent declines in blood lead concentration and reductions in such sources as leaded gasoline.

Plasma

Earlier data suggested that plasma lead does not vary across a broad range of total blood lead, but it is now accepted that plasma and erythrocyte lead are in equilibrium and that the plasma fraction of lead is stable up to a blood lead concentration of 50-60 μg/dL, at which the fraction increases (Cavalleri et al.,1978; DeSilva, 1981; Manton and Malloy, 1983; Manton and Cook, 1984).

The existence of an equilibrium of lead between plasma and erythrocytes indicates that some fraction of total erythrocyte lead can be shifted to plasma in responses to downward shifts from steady-state exposure. That accounts for the fast component of blood lead decay, which is commonly faster than that expected from erythrocyte turnover rates.

Plasma lead concentrations at steady state are extremely small, often less than 1% of blood lead concentration, and rarely above 1 μg/dL. Even at blood lead concentrations over 50-60 μg/dL, they go up to only a few micrograms per deciliter. That makes it unlikely that a typical clinical laboratory will routinely analyze plasma lead, owing to such complicating factors as ambient lead contamination and ready hemolysis of high-lead erythrocytes.

Teeth

Human deciduous teeth accumulate lead in substantial quantities over their embryonic and postnatal life, up to the time of shedding. Teeth are anatomically and metabolically diverse, and this affects lead toxicokinetics in the mineralizing matrix.

Like bone, teeth have long been recognized as relatively useful tissues for assessing biologic markers of long-term lead accumulation. Unlike bone, however, teeth irreversibly sequester lead (Cohen, 1970);

TABLE 4-4 Studies of Kinetic Behavior of Lead in Blood of Adults

Study Group and Exposure	Half-Life, days	Comments (References)
Experimental studies:		
Volunteers exposed to stable ^{204}Pb in diet	25	Short-term study, nonequilibrium kinetics for bone lead release (Rabinowitz et al., 1976)
Volunteers inhaling ^{203}Pb tracer aerosol	16	Short-term study, nonequilibrium kinetics for bone lead release (Chamberlain et al., 1978)
Volunteers inhaling cold lead aerosol at two concentrations	28 (10.9 $\mu g/m^3$) 26 (3.2 $\mu g/m^3$)	Short-term study, nonequilibrium kinetics for bone lead release (Griffin et al., 1975)
Epidemiologic studies:		
Men in England; low ambient exposure	120-180	Serial survey of blood lead after lowered exposure over longer time (Delves et al., 1984)
Women in English town; ambient exposure with tap water lead higher exposure	180	Lead plumbing changed; blood lead followed serially (Thomas et al., 1979)

Retired lead workers	2,044 (median)	Mean lead exposure, 23 years; broad range of half-lives (Schütz et al., 1987a)
Active lead workers, examined after removal to lower exposure	29 (median)	Half-life for short component of two-component curve, showing bone lead as do profiles of retired workers (Schütz et al., 1987a)
Lead workers; exposure reduced because of strike at plant	20-130	Broad range of short-term blood lead component (O'Flaherty et al., 1982)
Lead workers; exposure reduced because of medical removal for excessive exposure (≥ 60 μg/dL)	79-130	Broad range showing exposure history (Kang et al., 1983)
Lead-poisoned workers removed from active exposure for medical reasons	619 (median)	Major input from slower, bone-lead-based component (Hryhorczuk et al., 1985)
Retired lead workers examined with in vivo x-ray fluorescence for tibia lead		Blood lead in ex-workers primarily from bone resorption (Christoffersson et al., 1984)

and they are more accessible for study because they are shed (see, e.g., EPA, 1986a; Mushak, 1989).

Table 4-5 presents illustrative studies of lead distribution in human teeth and the potential utility of teeth as biologic markers of childhood lead exposure. The data in the table make it clear that lead deposition in teeth is complicated and is a function of age (Steenhout and Pourtois, 1981), region of tooth (e.g., Needleman and Shapiro, 1974; Grandjean et al., 1986), type of tooth (Mackie et al., 1977; Delves et al., 1982), and extent of exposure (EPA, 1986a; Mushak, 1989).

Secondary (circumpulpal) dentin accumulates the largest concentrations of lead and is most sensitive to the extent of lead uptake in other body compartments (through contact with blood lead), so it is probably best suited for examining even subtle exposure. It seems especially attractive to consider low-lead epidemiologic studies in tissues that are major accumulators of lead, as opposed to tissues that only accumulate lower concentrations. Teeth differ in lead content as a function of tooth type, concentrations being higher in incisors than in premolars; this kind of difference is related in part to the fractions of actively accumulating regions in each tooth type (see, e.g., Mackie et al., 1977).

Lead in shed teeth of children, however useful for revealing accumulation, reflects retrospective exposure over a fairly long period and one that encompasses peak sensitivity and peak exposure periods, i.e., at the age of 2-3 years. Hence, this measure remains of less use than some others as a basis for environmental intervention at specific times.

In vivo analysis of lead in teeth seems to have the virtue of providing information on current lead accumulation when used in tandem with serial measurements of blood lead. Shapiro and co-workers (1978) showed a moderate correlation between lead in teeth and blood lead as single measures. However, such an in vivo measure seems to offer little advantage over similar measurements of tibial sites.

Bone

The skeletal system accumulates lead from before birth until at least the sixth decade. As public-health concerns are increasingly shifted to smaller lead exposures, two aspects of bone lead rise in importance. The first is the increasing degree to which bone contributes lead to total blood lead concentration, especially during pregnancy and at later stages

TABLE 4-5 Studies of Kinetic Behavior of Lead in Human Teeth

Study Group and Exposure	Type of Tooth Measure	Comments (References)
U.S. children; high exposure	Circumpulpal dentin, shed teeth	U.S. urban children have higher tooth lead than controls (Needleman and Shapiro, 1974)
U.S. children; range of exposures	Whole tooth; various tooth types	Lead varies with tooth type (Mackie et al., 1977)
U.S. urban children; with higher lead exposures	Incisors in vivo, related to concurrent blood lead	Correlation of in vivo tooth lead with blood lead (single; $r = 0.5$) (Shapiro et al., 1978)
Danish children; much lower exposure than U.S. children	Circumpulpal dentin	Concentration varies with age and tooth type (Grandjean et al., 1986)
British children stratified by socioeconomic variables	Tooth crowns (shed tooth minus resorbed pulp)	Considerable variance with type and position in jaw (Delves et al., 1982)
Belgian children; variable lead exposure	Whole tooth	Normalizing for age gives better index of exposure (micrograms per gram per year) (Steenhout and Pourtois, 1981)

of life—e.g., in osteoporosis in postmenopausal women. The second is toxicokinetic and methodologic: the extent to which real-time monitoring of bone lead can be used to determine unsafe rates of body lead accumulation.

It used to be widely held that the human skeletal system provides a metabolically inert depository for lead and that the huge amounts of lead being sequestered were inconsequential for health-risk assessment. That confidence rested in part on the assumption that bone was kinetically homogeneous as a lead compartment, with a half-life long enough to forestall risk of ready transfer back to blood. Current evidence argues, however, that bone is both a set of compartments for lead deposition and a target for lead toxicity. The dual identity complicates bone lead kinetics when it is applied to long-term modeling. The mobility of bone lead to blood is important. Table 4-6 summarizes studies that helped to characterize bone lead as a potentially toxic fraction of whole-body lead in sensitive populations. (Some of the material overlaps that presented earlier on blood lead.)

Human bone appears to have at least two, and possibly three, kinetic-

TABLE 4-6 Studies of the Kinetic Behavior of Lead in Human Bone

Study Group and Exposure	Comments (References)
Swedish retired lead workers; 3-45 years of lead exposure	Bone lead adds approximately 65% to total blood lead in retirement; accounts for half-life of 5.6 years (Schütz et al., 1987a)
Swedish lead workers; work exposure variable; in vivo analysis compared with chelatable lead	Chelatable lead well correlated with trabecular, but not cortical bone (Schütz et al., 1987b)
Japanese lead workers at various ages	Chelatable lead is age-dependent, showing bone contribution (bone lead age-dependent) (Araki and Ushio, 1982)
U.S. urban high-risk children meeting test criteria for in vivo tibial lead vs. chelation test	Tibial (cortical) lead correlated with, and predictive (with blood lead) of chelatable lead (Rosen et al., 1989)

ally distinct lead compartments. Lead in trabecular (spongy) bone appears to be more mobile than lead lodged in cortical (compact) bone, and there appears also to be a fraction of bone lead in equilibrium with the lead in blood (see, e.g., Skerfving, 1988). Trabecular bone seems to be an important source of resorbed lead when high exposure is reduced, e.g., through removal for medical reasons, by retirement of lead workers, or in response to chelation in adults (Schütz et al., 1987a). In young children, in whom only cortical bone has been examined, lead appears to leave cortical bone (tibia) (Rosen et al., 1989).

In the aggregate, the information in Table 4-6 makes it clear that bone lead readily returns to blood in substantial proportions. Although the mobilization is most apparent in people with a history of occupational exposure, bone lead also appears to be a major contributor in older people with ambient exposures to lead. More important, it is clear that bone lead is constantly mobilized in young children as part of physiologic remodeling of bone in the growth process. In addition to having a smaller fraction of total body stores of lead in bone, young children continuously recycle lead from bone to blood and other nonosseous tissues in bone reformation that accompanies the growth process.

Milk

Milk is the primary dietary constituent for young infants. Although the concentration of lead in human milk and infant formulas is relatively low (about 1.7 μg/dL), the volume consumed is large and thereby constitutes the primary source of lead for young infants, amounting to a daily intake of up to 50 μg (Ryu et al., 1985). Lead content of breast-fed infants' milk correlates well with their blood lead concentrations until 6 months of age ($r = 0.42, p < 0.0003$) (Ryu et al., 1985), when infants begin to crawl and walk. At these times of infant and child development, milk lead content accounts for less than 10% of blood lead concentrations (Rabinowitz et al., 1985b).

Placenta

Some studies have found a close correlation between maternal and newborn blood lead concentrations (Rabinowitz and Needleman, 1982; Korpela et al., 1986), but changes in compartmentalization of lead between blood, soft tissues, and bone of both mother and fetus might well be nonlinear and affected by nutritional status, as well as by marked differences in body composition between the developing fetus and the pregnant woman. In fact, blood lead concentrations during pregnancy have been seen to decline, rise, or evidence no definite trend (Davis and Svendsgaard, 1987).

Birthweight, head circumference, and placental weight were reduced as a function of placental lead content in a group of 100 obstetrically normal infants (Ward et al., 1987). Moreover, in the Port Pirie study in Australia, preterm delivery, defined as birth before the thirty-seventh week of pregnancy, was significantly related to maternal blood lead concentrations at delivery in a dose-response manner (McMichael et al., 1986). The latter study was carried out in 831 pregnant women, and the data were assessed by multivariate techniques. It has also been reported that placental lead content increases with lead exposure (Roels et al., 1978; Khera et al., 1980; Mayer-Popken et al., 1986); and in a limited study of amniotic fluid lead concentrations, it was found that concentrations of lead at term (59.6 $\pm$ 8.3 ng/ml) were significantly higher than maternal blood lead (40.4 $\pm$ 18.2 ng/ml) and umbilical cord blood lead (37.1 $\pm$ 13.5 ng/ml) (Korpela et al., 1986). Moreover, amniotic fluid concentrations failed to correlate with maternal or cord blood concentrations. Measurements of bone lead and blood lead concentrations (in pregnant women throughout the course of pregnancy), assessments of amniotic fluid concentrations, and placental lead concentrations at term collectively hold promise for further characterizing the dynamics of maternal-fetal lead transfer.

Chelatable and Urinary Lead

Spontaneous excretion of lead in nonoccupationally exposed humans is a highly variable process that involves small concentrations of lead (EPA, 1986a). In view of the difficulty of analyzing lead at low concentrations in a complex matrix, urinary lead appears to have little utility for general screening.

In contrast, the plumburesis associated with lead mobilization provides what is considered the best measure of the potentially toxic fraction of the total body lead burden (see CDC, 1985, 1991; EPA, 1986a). On the basis of various in vitro experimental and epidemiologic studies (Chisolm and Barltrop, 1979; Piomelli et al., 1984; CDC, 1985; EPA, 1986a; Mushak, 1989), chelatable lead is assumed to be a chemical sample of both mobile body compartments—i.e., blood and soft tissues—as well as of subcompartments of bone.

BIOLOGIC MARKERS OF EFFECT

This section examines biologic markers of early subclinical effects of lead that have potential value in the quantitative assessment of human health risk. It serves as a bridge between human lead toxicology and the practice of laboratory screening in analytic toxicology for the evaluation of lead intoxication. Elucidation of biologic markers of effect also sheds light on mechanisms of toxic action. Methodologically, markers of effect are any measurable biochemical, physiologic, or other alteration from normal in humans that shows potential health impairment. The theoretical advantage of markers of effect over markers of exposure is that markers of effect reflect actual biologic responses of the body. A practical advantage in many instances is that markers of effect are relatively independent of the vicissitudes of lead measurement, particularly the contamination of samples with lead.

If a biologic marker is a robust early perturbation in response to low-dose exposure, it might be found in most of the target population. It is important that such an early perturbation not be likely itself to constitute a bona fide adverse effect and that it be useful in avoiding exposure sufficient to produce adverse effects in other organ systems. Given present knowledge, a biologic marker should reliably operate at a blood lead concentration below the 10 μg/dL associated with population IQ decrement, neurobehavioral changes, and deficits in growth indexes in a discernible fraction of young children. Some (e.g., Friberg, 1985) have even argued that markers of effect themselves indicate that it is already too late for monitoring to prevent any toxicologic perturbation at all.

Biologic markers of lead's effects are often confined in usefulness to a range of body burden of lead. As acceptable magnitudes of lead exposure have been reduced, it has been necessary to re-evaluate the relevance of biologic markers of effect, as is the case with biologic markers based on lead's disturbances of heme synthesis.

This section deals with the epidemiologic utility of established markers of effect, including brief statements on the underlying toxicology and pathology. The markers are discussed in terms of their current usefulness and reliability as the definition of "safe" lead exposure continues to change. The section also discusses the use of markers for elucidating mechanisms of toxic action.

Markers Based on Disturbance of Heme Synthesis

Lead affects the biologic synthesis of heme at various steps in a number of important organ systems (see Chapter 2, especially Fig. 2-5). Overall toxicity in the synthetic pathway at moderate exposures is attributable to effects on three enzymes, although others can be affected at larger exposures. The three effects are stimulation of the activity of delta-ALA synthetase (delta-ALA-S), the mitochondrial enzyme that mediates the rate-limiting step in heme formation; inhibition of the activity of the cytosolic enzyme porphobilinogen synthetase (PBG-S or delta-ALA-D); and inhibition of the activity of the mitochondrial enzyme ferrochelatase or inhibition of intramitochondrial movement of iron to the ferrochelatase site.

The steps in heme synthesis are not uniformly sensitive to lead, nor are they all equally useful in development of biologic markers of effect; only some steps are useful this way. Some of the relevant characteristics of biologic markers based on disturbance of heme synthesis are presented in Table 4-7.

Inhibition of the activity of the enzyme delta-ALA-D occurs at a very low body lead burden, indexed as blood lead; the threshold of this effect is 5 μg/dL or even lower (Chisolm et al., 1985; EPA, 1986a). Hernberg and Nikkanen (1970) produced data that allow an estimate of 50% inhibition at a blood lead concentration of 16 μg/dL. Thus, at current exposures in the United States, many people would be expected to have measurable inhibition of the enzyme.

The enzyme is retained in the mature erythrocyte with a function that is vestigial compared with its role in blood-forming and other tissues.

The erythrocyte enzyme's response to lead is similar to its response in other tissues at high lead concentrations (Secchi et al., 1974; Dieter and Finley, 1979; EPA, 1986a), but tissue activity at low concentrations is unknown.

Lead directly affects delta-ALA-D activity by active-site inhibition through thiol-site binding (e.g., Finelli et al., 1975). That behavior produces two problems: one is related to diagnostic utility, in that direct measurement of lead is equivalent to measurement of delta-ALA-D enzyme activity, and vice versa; the second is methodologic, in that measurement of delta-ALA-D activity is affected by lead contamination of the sample just as direct blood lead measurement is. In addition, enzyme activity is affected by zinc contamination: zinc offsets lead inhibition and produces inaccurate results.

The distribution of delta-ALA-D in sensitive populations is genetically polymorphic and occurs as three phenotypes (Doss et al., 1979; Battistuzzi et al., 1981; Astrin et al., 1987). Consequently, delta-ALA-D-activity screening in tandem with phenotype identification (Astrin et al., 1987) distinguishes subjects who are genetically most susceptible at a given blood lead concentration and those who merit maximal protection from exposure.

With increasing inhibition of delta-ALA-D activity, delta-ALA accumulates in the body and eventually in urine. The threshold for urinary ALA accumulation (as blood lead concentration) can be up to 40 μg/dL for workers and children, depending on measurement method (NRC, 1972; Chisolm et al., 1976; Meredith et al., 1978; EPA, 1986a; Okayama et al., 1989).

Not only does urinary ALA accumulation have a high threshold in terms of magnitude of lead exposure, but its utility in low-exposure screening is generally assumed to be valid for groups, rather than for individuals, such as lead workers (e.g., Roels et al., 1975; Alessio et al., 1979). Statistically, the sensitivity and the specificity of the method are such that high predictability (minimal false negatives or false positives) exists only at high blood lead concentrations (e.g., Okayama et al., 1989); that rules out its use in young children and pregnant women who receive low-dose lead exposures. Reports of screening of high-risk children with colorimetric measurement of urinary ALA indicate poor correlation with blood lead concentrations (Blanksma et al., 1970; Specter et al., 1971; Chisolm et al., 1976).

TABLE 4-7 Heme-Synthesis Disturbances and Effect Markers

Lead Effect	Result	Marker Threshold, Lead Concentration, µg/dL	Comments
Inhibition of delta-ALA-D (PBG-S) activity	Accumulation of ALA in tissues and urine	5	Sensitive for current population blood lead concentrations; problematic relation to tissue effects
Feedback stimulation of delta-ALA-S activity	Minor contribution to total ALA in urine	40	Not a feasible marker
Accumulation of urinary ALA	--	20-40[a]	Useful for population screening; limited in individual predictability; not useful for childhood screening
Inhibition of heme formation from protoporphyrin IX	Accumulation of erythrocyte protoporphyrin in blood	15-20 (children) 25-30 (adults)	Most common screening marker for children and workers; basis of risk scheme, with blood lead, used by CDC in recent advisories, but not a part of the CDC 1991 statement

| Impaired use of coproporphyrin | Accumulation of coproporphyrin in urine | 40 | Reflects continuing toxicity; supplanted in popularity by erythrocyte protoporphyrin measurement |

[a]Depends on method of measurement (Chapter 5).

The aspect of heme-synthesis disturbances by lead that has been most widely exploited as a biologic marker of early effect has been the accumulation of the heme precursor erythrocyte protoporphyrin IX or zinc protoporphyrin (EP or ZPP) in blood of children and in some adult populations. EP accumulates in response to lead-related inhibition of the activity of the intramitochondrial enzyme ferrochelatase or lead-related impairment of intramitochondrial iron transport (Chisolm and Barltrop, 1979; CDC, 1985; EPA, 1986a; Moore et al., 1987). EP increase therefore indicates a generalized mitochondrial toxic response. It accumulates only in newly formed erythrocytes during the active lead-exposure period, and it takes weeks after onset of exposure for it to show up. It remains increased after lead exposure ceases, and it decreases in proportion to the turnover rate of the human mature erythrocyte, i.e., about 0.8%/day in the absence of decreased cell survival.

EP accumulation is exponentially and directly correlated with blood lead in children (Chisolm and Barltrop, 1979; Piomelli et al., 1982, 1984; Hammond et al., 1985) and in adults (Grandjean and Lintrup, 1978; Lilis et al., 1978; Valentine et al., 1982; Alessio, 1988). The population threshold of blood lead concentration for a lead-associated EP response is 15-20 μg/dL in children (Piomelli et al., 1982; Hammond et al., 1985) and 25-30 μg/dL in lead workers (Grandjean and Lintrup, 1978).

The utility of EP accumulation in a rapid and cost-effective screening procedure for high-risk children in the United States was recognized early; it was so attractive for screening at the high blood lead concentrations common in the early 1970s that it became part of the screening method advanced by the U.S. Public Health Service in 1975 (CDC, 1975). In its 1975 statement, the Centers for Disease Control and Prevention linked EP of 60 μg/dL of whole blood to later measurement of lead in venous blood. A combination of a blood lead of 30 μg/dL or higher and an EP of 60 μg/dL or higher was taken as evidence of lead toxicity. In 1978, the combination was modified to lead of 30 μg/dL and EP of 50 μg/dL or higher (CDC, 1978). The new CDC statement (CDC, 1991), however, makes it clear that use of EP is not practical or useful for low blood lead concentration in portions of the multitiered approach now being recommended.

Lead-associated EP increase is similar in hematologic result to iron deficiency, so the use of blood EP as a lead-specific marker must take into account the relative risk of iron deficiency or the actual iron status of the people being screened (CDC, 1985; Mahaffey and Annest, 1986; Marcus and Schwartz, 1987; Piomelli et al., 1987). But it has to be kept in mind that the rare genetic disorder erythropoietic protoporphyria produces high concentrations of EP that are discordant with blood lead concentrations.

Increase in EP concentration, as a measure of lead exposure, differs from increase in urinary coproporphyrin (CP), a heme precursor that was traditionally used as a marker of childhood and worker exposure to lead before the heavy use and popularity of EP measurement. Urinary CP responds only to active lead exposure and intoxication and is a measure that reflects current exposures (Piomelli and Graziano, 1980). The threshold for urinary CP increase appears to parallel that of urinary ALA, being about 40 µg/dL in lead in blood.

The apparent interference of lead in the kidney 1-hydroxylase system, which uses heme, is discussed with respect to vitamin D later in this chapter.

Markers of Other Biologic Systems

Biologic markers that are based on biologic systems other than heme synthesis are presented in Table 4-8.

Lead intoxication produces impairments in blood-forming tissue other than those directly involved in heme synthesis. Lead strongly inhibits an enzyme central in erythropoietic pyrimidine metabolism, pyrimidine-5'-nucleotidase (Py-5'-N). The enzyme catalyzes the hydrolysis of pyrimidine nucleotides from the degradation of ribosomal RNA fragments in maturing erythrocytes. Inhibition leads to accumulation of the nucleotides, and ribosomal catabolism is retarded (Paglia and Valentine, 1975; Angle and McIntire, 1978; Buc and Kaplan, 1978); at high lead exposures, inhibition is severe enough to produce basophilic stippling from undegraded fragments.

The blood-lead threshold for this effect, based on lead-exposed children, appears to be around 10 µg/dL (Angle et al., 1982; Cook et al., 1986, 1987). Sensitivity for predicting a blood lead concentration equal to or greater than 40 µg/dL is 80% (using an enzyme activity mean below 2 standard deviations (SDs)); specificity, as percentage of

TABLE 4-8 Non-Heme-Synthesis Markers of Effect of Lead Exposure

Lead Effect	Result	Marker Threshold, μg/dL	Comments
Inhibition of Py-5'-N activity in erythrocytes	Accumulation of ribosomal fragments in reticulocytes	5-10	Analogous to role of ALA-D activity inhibition; quite sensitive; linkage to adverse effects questionable
Inhibition of Na^+,K^+-ATPase in erythrocyte membrane	Potassium loss and net sodium gain in cells; altered cell survival	Not established; in workers, limited correlation with blood lead	Studies in lead workers; direct measure of lead's presence; subject to contamination
Inhibited hydroxylation of 25-OH-vitamin D	Reduction in hormonal metabolite 1,25-$(OH)_2$-vitamin D	10-15	Important health effect, but not appropriate for use as marker

children with an adequate enzyme activity who had blood lead less than 40 µg/dL, was 96 (Cook et al., 1987). The status of this marker for its relevance to lead toxicity is analogous to that of delta-ALA-D, noted above. Like delta-ALA-D, it reflects the presence of biologically active lead at a site not readily probed directly. At some point in increasing lead exposure, lead-related inhibition of the enzyme will produce an accumulation of pyrimidine metabolites. Py-5'-N activity is also quite low in a genetic disorder that produces a hemolytic anemia due to such inhibition and ribosomal fragment buildup (Valentine and Paglia, 1980).

Consequently, people with this phenotype, which is rare, are extremely sensitive to lead exposure and merit both identification and added protection from lead exposure.

Lead exposure results in inhibition of erythrocyte membrane Na^+,K^+-ATPase, which mediates the control of intracellular movement of both these physiologically crucial ions (Raghavan et al., 1981). Inhibition produces loss of cellular K^+, but does not disturb input of Na^+; it produces cell shrinkage, a net increase in sodium concentration, and increased fragility and lysis.

This marker of membrane ATPase has been examined quantitatively only in lead workers (Secchi et al., 1968; Raghavan et al., 1981) in whom it appears that the inhibition correlates well with membrane lead concentrations, but poorly with total blood lead. Its broad applicability to screening of lead workers and other populations is problematic.

In young children, increases in blood lead are associated with disturbances in vitamin D function, notably formation of the hormonal metabolite 1,25-$(OH)_2$-vitamin D, which is crucial for a wide variety of functions (see Chapter 2). The threshold for reduction of concentrations of 1,25-$(OH)_2$-vitamin D in terms of blood lead appears to lie at blood lead concentrations of 10-15 µg/dL (Rosen et al., 1980; Mahaffey et al., 1982b); the effect on vitamin D function is uniformly distributed across the range of blood lead concentrations.

Reductions in plasma concentrations of the hormonal metabolite are not suitable for use as biologic markers of effect. The reductions in this metabolite and some other early effects are now known to occur in some people at quite low concentrations of blood lead (including cognitive and other neurobehavioral end points) and are, in fact, adverse effects, given the crucial function of this hormone in the body, and one must therefore use other more sensitive effect markers before decrements in circulating concentrations of 1,25-$(OH)_2$-vitamin D would be detected.

Relevance of Current Markers of Effect for Low-Dose Exposures

The utility of commonly used biologic markers of effect is related to the range of body lead burden of concern. Summary comments regarding these are provided in Table 4-9.

Use of increase in EP in lead screening of high-risk populations

TABLE 4-9 Markers of Effect Relative to Low-Dose Lead Exposure

Marker	Threshold, μg/dL	Effectiveness
Activity of ALA-D (PBG-S) inhibition	5	Sensitivity at low exposure; relevance to tissue effects questionable; useful for phenotype sensitivity to lead toxicity
EP increase	15-20 (children) 25-30 (adults)	Yields too many false-negative results at blood lead around 25 μg/dL; no correlation at blood lead of 10-15 μg/dL
Urinary ALA increase	Up to 40	Not useful at blood lead of 10-15 μg/dL
Urinary CP increase	40	Not useful at blood lead of 10-15 μg/dL
Py-5'-N activity inhibition	5-10	Quite sensitive at blood lead of 10-15 μg/dL; relevance to toxic effects questionable

appears to yield an increasingly unacceptable rate of false-negative results as population blood concentrations decline and as the guideline concentrations for what is acceptable exposure also fall (see Chapter 2). Mahaffey and Annest (1986) showed that at the relatively high blood lead concentration of 30 µg/dL, the false-negative rate for blood lead and EP in the NHANES II survey population approached 50%. As seen in Chapter 5, such false-negative rates are lower, but are still high in high-risk child populations like those in Chicago. One would expect the rate to be even higher at lower blood lead concentrations, as seen in analysis of the Chicago screening population cited.

Such measures as urinary ALA and CP apply mainly to relatively high lead exposures, having thresholds of about 40 µg/dL. They would not be of much use at current concentrations of concern, around 10 µg/dL.

Inhibition of activity of both delta-ALA-D and Py-5'-N is significant at blood lead concentrations of 10 µg/dL and less. Consequently, they would technically be useful markers; but they appear to offer little advantage over lead measurement itself, because the presence of erythrocyte lead produces continuing inhibition, and contaminating lead (and zinc in the case of delta-ALA-D) would also affect the measurements of enzyme activities. In the case of delta-ALA-D, and perhaps that of Py-5'-N, use of these markers in tandem with assessment of genetic phenotype would identify subjects at great risk of hematotoxicity even at low doses of exposure. That would be especially important in children homozygous for the ALA-D-2 allele (Astrin et al., 1987), but also important in heterozygous children. A significant number of such phenotypes are seen in traditionally high-risk populations (e.g., Doss et al., 1979; Rogan et al., 1986; Astrin et al., 1987).

Potential Markers of Effect

Despite the limited utility of many of the conventional biologic markers of effect either at high exposures for such organs as the kidney or at increasingly lower general exposures, it is still of interest to consider available information on some candidates for biologic markers. These are classified as enzymes, binding proteins, and metabolites. A summary of the potential markers of effect is given in Table 4-10.

TABLE 4-10 Potentially Useful Markers of Effect of Lead Exposure

Marker	Threshold	Comments
Enzymes		
Urinary N-acetyl-beta-D-glucosaminidase	Approximately 60 µg/dL	Most sensitive measure of tubular injury by lead so far
Erythrocyte nicotinamide adenine dinucleotide (NAD) in exposed subjects and in vitro	Not measured, but lower than that for Py-5'-N	Reflects damage to formation and function of NAD; suggests that genetic susceptibility to NAD reduction can be exacerbated by lead exposure
Lead-Binding Proteins		
Erythrocyte low-weight binding proteins in association with lead toxicity	Protein low in lead workers with overt toxicity	Serves a protective function analogous to suggested function in animal kidney
Metabolites		
beta-Isobutyric acid (β-IBA) in urine	Increased threefold at blood lead approximately 60 µg/dL; no threshold calculated	Suggests that lead is damaging DNA function and formation via increased thymidine degradation to β-IBA

Enzyme Systems

As noted in Chapter 2, lead is now widely known to induce both glomerular and tubular injury, at least in adults occupationally exposed to lead. In most instances, however, the thresholds for such injury appear to be high and presumably of interest mainly in occupational screening at current OSHA guidelines. Schaller et al. (1980) and Buchet et al. (1980) reported that no discernible kidney dysfunction was observable below 60 μg/dL (Buchet et al., 1980) or over the range 50-85 μg/dL (Schaller et al., 1980). In common with the rest of the field of occupational or environmental metal nephrotoxicity, however, the indicators of what constitutes early kidney injury due to lead have been relatively insensitive until recently.

The lower range of lead-induced nephrotoxicity was examined by Meyer et al. (1984) and Verschoor and co-workers (1987). Their main index for low-concentration lead effect was urinary excretion of the tubule cell lysosomal enzyme N-acetyl-beta-D-glucosaminidase (NAG). This enzyme in urine might well be a promising general marker of early metal-induced kidney damage, in that various workers found it to be more sensitive than beta-2-microglobulin in assessing early cadmium-induced tubular dysfunction in workers (e.g., Chia et al., 1989; Kawada et al., 1989; Mueller et al., 1989).

In their worker group, Verschoor et al. (1987) found that tubular, rather than glomerular, indexes are affected earliest and that urinary NAG concentrations were increased at blood lead below 60 μg/dL and showed a significant correlation with EP.

Nicotinamide adenine dinucleotide (NAD) synthetase (NAD-S) catalyzes the final step in the Preiss-Handler pathway for biologic synthesis of the metabolically common coenzyme NAD (Preiss and Handler, 1958), and it is associated with an end product whose synthesis is impaired in a number of genetic disorders, including thalassemia (Zerez and Tanaka, 1989) and sickle-cell disease (Zerez et al., 1988).

Zerez et al. (1990) found that, in erythrocytes from lead-exposed subjects or treated with lead in vitro, NAD-S activity is obliterated at lead concentrations at which Py-5'-N activity, itself a sensitive measure of lead exposure, still retains 50-70% of activity. The small group of lead-exposed subjects ranged in blood lead from 34 to 72 μg/dL.

Lead-Binding Proteins

The presence of proteins with an avidity for lead in kidney and brain was discussed in Chapter 2. In the erythrocytes of variably exposed lead workers, there is a lead-binding protein that appears to be inversely linked to clinical manifestations of occupational lead intoxication; i.e., the higher the erythrocyte concentration, the more resistant the worker appears to be to overt poisoning (Raghavan et al., 1980, 1981; Lolin and O'Gorman, 1986). The protein's function is reminiscent of kidney cytosol lead-binding protein (e.g., Oskarsson et al., 1982; Goering and Fowler, 1985).

Lolin and O'Gorman (1988) measured the protein in lead workers with various degrees of lead exposure. The protein was quantitated as two peaks, which suggested a heterogeneous protein. It was present in erythrocytes from all lead workers, but absent from controls. The threshold for induction of the protein was about 38 μg/dL or less; that indicates some utility for monitoring in occupational exposures. Equally important, concentrations of the protein are significantly lower in people with clinical toxicity. Furthermore, those workers found that the concentration of the erythrocyte lead-binding protein was related to intensity of exposure (past and present), not its duration. That is also typical of inducible proteins.

Lolin and O'Gorman have postulated that the protein is protective in its function and would play a special role in subjects who are particularly susceptible to lead's effect on, e.g., ALA-D activity. They had found earlier (Lolin and O'Gorman, 1986) that there is a type of ALA-D activity inhibition not seen in environmental or nonoccupational exposures. Such protection targeted to preservation of ALA-D activity is consistent with the findings of Oskarsson et al. (1982) and Goering and Fowler (1985) that rat kidney ALA-D activity is notably resistant to lead and that this is due to the presence of a kidney cytosolic lead-binding protein. The analytic data on both worker erythrocyte and rat kidney cytosol protein structure suggest metallothionein, as noted by Goering and Fowler (1985) and Lolin and O'Gorman (1988), but further work is required to establish this fact.

Metabolites

beta-Aminoisobutyric acid (beta-AIB) is a normal degradation product of thymidine, a constituent of DNA. Unlike typical amino acids, it is actively secreted as a catabolic metabolite via the tubule. It is normally excreted at low rates in humans not exposed to lead (6 nmol/μmol of creatinine). Farkas and co-workers (1987) examined the concentrations of this metabolite in urine of workers occupationally exposed to lead and in marmoset monkeys experimentally exposed via tap water. In workers with a mean blood lead concentration of 64 μg/dL and a mean EP of 117 μg/dL, there was a tripling of urinary output of beta-AIB. No threshold was determined for excretion beyond the normal range. In monkeys, there was a dose-dependent increase in urinary excretion. Given the fact of beta-AIB's handling by the kidney, the increase in the metabolite is a marker more of DNA damage through increased degradation of thymidine to beta-AIB.

Identification of Toxicity Mechanisms

Markers of effect not only are useful in the screening of high-risk populations, but also help to establish the various molecular and cellular mechanisms by which lead imparts multiorgan toxicity in those high-risk populations.

Inhibition of 1,25-Dihydroxyvitamin D Formation

As noted elsewhere, lead exposure is associated with reduced blood concentrations of the hormonal metabolite of vitamin D, 1,25-$(OH)_2$-vitamin D. Such reductions with blood lead concentrations of 33-55 μg/dL, furthermore, rival those seen in several disease states (Rosen et al., 1980; Mahaffey et al., 1982b; Rosen and Chesney, 1983). Consequently, reductions in this hormone at lower lead exposures are signaling early metabolic disturbance. Reduced concentrations of the hormone indicate that lead has two mechanisms of adverse effect that potentially can operate in high-risk populations. The first concerns the toxic consequences of disturbance in the hormone-calcium relationship, and the second concerns the many roles played by 1,25-$(OH)_2$-vitamin D beyond regulatory control of calcium function.

A major mechanism of cellular lead toxicity appears to be interference in calcium homeostasis and function (Chapter 2). Such interference occurs either directly, via lead-calcium interactions in the cell, or through impaired function of calcium as a second messenger due to disturbed regulation by 1,25-$(OH)_2$-vitamin D (Rasmussen, 1986a,b; Pounds and Rosen, 1988). It implies a risk of impaired handling of vesicular intestinal calcium and intracellular calcium in bone cells. Calcium-based effects are broadly distributed as to tissue and system sites, including the vascular system and developing neural and bone tissue.

As summarized in Table 2-3, other physiologic functions potentially can be altered by reduced concentrations of 1,25-$(OH)_2$-vitamin D. They include parathyroid phospholipid metabolism, cyclic GMP production in skin fibroblasts, renal phosphate reabsorption, and differentiation and proliferation of diverse cell types. In addition, the division, communication, and cytostructural organization of many cell types are affected.

Impairment of Heme Synthesis

Heme is a prosthetic group for many functional proteins involved in cell function and survival, and its formation is obligatory for cellular functions in many tissues, especially blood-forming tissue, muscle, kidney, liver, and brain (EPA, 1986a). Evidence of an effect of lead on heme formation would constitute a far-reaching mechanistic clue to lead toxicity.

Heme formation is an intramitochondrial process. Its inhibition, whether by inhibition of the intramitochondrial enzyme ferrochelatase or by impairment of intramitochondrial delivery of the iron atom to protoporphyrin, can be considered a marker of generalized mitochondrial toxicity of lead in heme formation for a large number of cell and tissue types.

BIOLOGIC MARKERS OF SUSCEPTIBILITY

One factor that can enhance susceptibility to lead toxicity is nutritional status. Various nutritional factors have been shown experimentally to influence absorption and tissue concentrations of lead (Mahaffey, 1985). Although experimental diets can be manipulated to create a wide range of nutritional problems, factors of clinical importance are far more restricted. Those considered of greatest importance to young children, as well as many adults, are total food intake, frequency of food intake, and dietary intake of some trace minerals, notably calcium and iron (Mahaffey, 1985).

Fasting adults have been reported to absorb a substantially greater fraction of dietary lead than nonfasting adults (Rabinowitz et al., 1980; Heard and Chamberlain, 1982). Comparable data on the effects of fasting on lead absorption by children and young nonhuman primates are not available.

To assess the role of dietary calcium in absorption and retention of environmental lead, it is essential to recognize that effects reflect both acute and long-term variation in dietary calcium intake. Numerous studies with experimental animals fed diets low in calcium have established that calcium deficiency increases both tissue retention and toxicity of lead (Mahaffey et al., 1981). Long-term calcium deficiency produces physiologic adaptive mechanisms (Norman, 1990), including increased concentrations of various binding proteins and stimulation of the endocrine and regulatory systems that regulate the concentration of ionized calcium. Parathyroid hormone and $1,25\text{-}(OH)_2$-vitamin D are critical to regulatory control of calcium.

Physiologic controls that react to changes in dietary calcium also affect the biokinetics of lead. Generally, calcium deficiency increases lead toxicity (Mahaffey et al., 1973), but it is not clear that the increase in toxicity occurs predominantly because of physical competition between calcium and lead for absorption. The mechanisms that produce changes in lead absorption as a function of calcium status are not well documented.

Humans can adapt to a wide range of iron requirements and intakes (Cook, 1990). There are two basic principles of adaptation to differences in iron intake: Regulation of iron absorption is achieved by the gastrointestinal mucosa, and fractional absorption depends directly on the metabolic need for iron. How the gastrointestinal tract achieves adaptation to change in dietary iron intake and change in iron requirement remains an unanswered question in iron metabolism (Cook, 1990).

Iron deficiency increases tissue deposition and toxicity of lead (Mahaffey-Six and Goyer, 1972). Ragan (1977) demonstrated sixfold increases in tissue lead in rats when body iron stores were reduced, but before iron deficiency developed. The influence of iron status on lead absorption has been investigated in adult humans, but with different results (Watson et al., 1980, 1986; Flanagan et al., 1982). It is not clear whether the difference reflects the severity of iron deficiency, differences in analytic approach, or some other undefined factors. A high prevalence of iron deficiency occurs in infants, children, and adolescents because of the need to expand the body's iron pool for growth. Women have higher iron requirements because of menstrual blood losses. The iron requirement of a normal pregnancy is approximately 500 mg, which is distributed to the fetus, placenta, and expanded maternal erythrocyte mass.

It is critical to recognize that the groups of people who have the highest environmental lead exposures are also at greatest risk of iron deficiency (Mahaffey and Annest, 1986). The greatest impact of iron deficiency is in young children, who develop defects in attention span that lead to learning and problem-solving difficulties (Lozoff et al., 1985, 1987; Pollitt et al., 1986, 1989). Data from a longitudinal prospective study in Yugoslavia (Graziano et al., 1990) show the combined effects of lead exposure and iron deficiency among pregnant women, infants, and children. Adverse effects of both conditions on the neurobehavioral and hematopoietic systems were found.

The importance of iron status for low-income families has been recognized for decades. In the early 1970s, a special program for low-income women, infants, and children was begun in the United States. Extensive evaluation of the impact of a program of nutritional supplementation for women, infants, and young children has been reported by Rush et al. (1988a,b).

Children's blood lead concentrations tend to be associated with families' provision of intellectual and sociologic support (Milar et al., 1980; Hunt et al., 1982; Stark et al., 1982; Dietrich et al., 1985). In one study, the scores of mothers on three Home Observation for Measurement of the Environment scales were significantly associated with cumulative blood lead concentrations in infants: maternal involvement with child, provision of appropriate play materials, and emotional and verbal responsivity of the mother (Dietrich et al., 1985). Scores on such scales are loosely associated with socioeconomic status, although the prevalence of these risk factors varies substantially within all social strata (Pearson and Dietrich, 1985). The association between poor maternal support for the child and blood lead remains after other factors are controlled for (Bornschein et al., 1985).

One intrinsic, biologic factor of growing interest, but that has received little attention, is the distribution in sensitive populations of genetic susceptibilities that potentiate health risk. Few specific biologic markers of susceptibility for lead have been identified.

The nature of lead effects and their genetic potentiation (or attenuation) must be understood both qualitatively and quantitatively. The relative distribution of genetically susceptible segments of the population is of main concern when these segments suffer substantial lead exposure. For example, lead exposure and the hepatoporphyric genetic disorder acute intermittent porphyria both produce accumulation of potentially neurotoxic ALA in plasma and urine (e.g., EPA, 1986a) and show some neuropsychiatric responses. A second genetic disorder, that associated with ALA-D deficiency, might well be a special problem for children at high risk for lead exposure in urban areas of the United States (Astrin et al., 1987). According to available data, any increased, genetically based susceptibility to lead exposure or its adverse effects is based mainly on lead's effects on heme biosynthesis and erythropoiesis.

There is genetic polymorphism for the heme-pathway enzyme ALA-D in human populations. That has been recognized for some time (e.g., Granick et al., 1973), but the molecular genetic basis of the phenomenon has now been described (Battistuzzi et al., 1981). Potluri et al. (1987) identified the gene site at chromosome 9q34. Two common alleles, ALAD-1 and the deficiency ALAD-2, with frequencies of 0.9 and 0.1 in Europe-based populations, give rise to three phenotypes: 1-1, 1-2, and 2-2. Heterozygotic (1-2) people have ALA-D activity of approximately 50% of normal, and severely deficient homozygotic (2-2) people have activity of approximately 2% of normal (Doss et al., 1979). Doss et al. (1982) reported that workers with moderate workplace contact with lead but manifest lead poisoning in the form of high erythrocyte protoporphyrin concentrations were found to be heterozygotic for ALA-D deficiency, and Doss and Muller (1982) described an acute lead-toxicity response in a person with ALA-D deficiency and moderate lead exposure. Ziemsen et al. (1986) reported that the fraction of lead workers with high blood lead increased as ALAD-2 phenotypy increased. Several recent studies have documented that children apparently heterozygotic or homozygotic for ALAD-2 can be susceptible to lead effects. Rogan et al. (1986) examined a group of children in a large lead-screening program and found that, independently of blood lead, children with ALA-D deficiency also had significantly high EP; results of further testing suggested a problem with the amount of the enzyme, rather than a biochemically defective form. Astrin et al. (1987) have, however, found that ALAD-2 in a large sample of lead-screened children in New York City was correlated with the relative frequency of high blood lead (over 30 μg/dL).

In acute intermittent porphyria, there is significant inhibition of the activity of the enzyme porphobilinogen deaminase, which mediates the conversion of porphobilinogen to uroporphyrinogen I. That leads to accumulation of high concentrations of ALA in urine and other body fluids (Goldberg and Moore, 1980; Moore et al., 1987). In both overt lead poisoning and attacks of acute intermittent porphyria, there are pronounced gastrointestinal, psychomotor, and cardiovascular responses, which have led to suggestions that lead works its adverse neurologic effects through direct action, as well as through the heme pathway (Silbergeld et al., 1982; EPA, 1986a). Conclusive evidence that lead significantly exerts neurotoxicity through the heme pathway, via excessive production of neurotoxic ALA (EPA, 1986a), has not been forthcoming. Studies directed to the hypothesis have entailed large exposures to lead, and it is not clear that low-dose lead exposure would effectively synergize neurologic manifestations of the attack stage of acute intermittent porphyria.

Increased lead exposure affects the liver in various ways (EPA, 1986a). One site of toxicity involves impairment of hepatic biotransformations of endogenous metabolites (e.g., Saenger et al., 1984) and impairment of the detoxification and activation of drugs and other xenobiotic substances via effects on P-450 mixed-function oxidase (Alvares et al., 1976; Meredith et al., 1978) and perhaps the carcinogen-activating P-448 complex. Evidence of genetically differentiated hepatic biotransformation and biodegradation capacity in the P-450 and P-448 systems has been reviewed by Parke (1987). It is already known that genetic diversity has an impact on biotransformation of various drugs (Parke, 1987). Animal systems show genetic polymorphism in P-450—P-448 systems, but the requisite human studies of the underlying molecular genetics remain to be done.

SUMMARY

The absorption, distribution, retention and excretion of lead in sensitive populations from various sources affect both the biologic monitoring of lead exposure and toxic outcomes. The literature dealing with the quantitative aspects of lead toxicokinetics is extensive and supports a number of conclusions.

Inorganic lead is absorbed principally from the lungs and gastrointestinal tract of humans. About 30-50% of inhaled lead is absorbed from the lower respiratory tract in adults; the proportion is greater in children. Lead in water is absorbed variably: in adults, 10-15% is absorbed after consumption of blood; in children, approximately 50% is absorbed after consumption of food. Under fasting or semifasting conditions, rates of absorption rise considerably for adults and probably for children.

Lead is absorbed into plasma; with steady-state exposure, redistribution of 99% or even more to the erythrocytes occurs. Newly absorbed plasma lead is distributed metabolically to blood, soft tissue, and various compartments of bone, where longer-term storage occurs. It is critical to recognize that skeletal lead stores are continuously remobilized as part of the physiologic remodeling of bone that accompanies the growth process.

Skeletal accumulation of lead occurs through life until the sixth or seventh decade, when lead loss from bone occurs. Among adults, lead is released from bone in response to reductions in exposure or to metabolic stresses and alterations in the skeletal system. In people not occupationally exposed, up to approximately 200 mg of lead can accumulate. In lead workers, much more accumulates. Reversal of accumulation is associated with either dietary changes or, especially in postmenopausal women, bone-mineral homeostatic changes. The latter contribute to the blood lead burden and complicate lead toxicokinetic compartmental analysis.

Lead is excreted by the kidneys and gastrointestinal tract (biliary clearance occurs). Urine is the dominant route in adults. Spontaneous lead excretion is variable and is affected by biologic processes that complicate analysis, but plumburesis associated with chelation probing or chelation therapy is extensive and diagnostically useful in lead poisoning.

Whole blood lead is the most popular and most useful indicator of lead exposure in acute and subacute poisoning. Blood lead measurement is also the most widely used measure of chronic lead exposure.

Blood lead reflects relatively recent exposure in young children who were not excessively and not chronically exposed in their earliest years; but heavily exposed children and adults have a blood lead concentration at any time that integrates recent and older exposures. With respect to quantitative inputs to a given blood lead content, recent contributions account for the major fraction in both children and adults. Recent input is associated with a quick component, and accumulated lead input is reflected in a much slower component, with a longer half-life (e.g., Schütz et al., 1987a).

Older exposures come into play via lead release to blood. One result of this phenomenon is a life-long legacy of earlier exposures. The policy and biomedical consequences are obviously important. For example, one must take account of the extent of such contributions to blood lead concentrations in planning the extent of control actions and concomitant population responses to lead regulatory actions.

At low-dose lead exposures, which induce effects that are of increasing concern, blood lead concentrations around 10 μg/dL or less must be monitored in various populations. About that concentration are distributions of blood lead concentrations—an aggregate variance, which is a complex integral of exposure, interindividual variance, and the scatter associated with laboratory methods. Of those factors, the magnitude of exposure and the quality of method refinements would be the most responsive to attempts at minimization. Individual differences in lead toxicokinetics are intrinsic to populations and not readily amenable to control.

Plasma lead, if it were amenable to clean collection and measurement, would have considerable interpretive value for assessing population lead exposures. Plasma, of course, transports lead to target tissues and major repositories. Contamination and loss problems are formidable for ready analysis of plasma in the typical clinical or epidemiologic research laboratory. Not only is contamination a problem for plasma analyses, but small rates of erythrocyte hemolysis would increase error. A hemolysis rate of 1% would double the plasma lead content, assuming a 1% equilibrium lead fraction in plasma. Furthermore, we still need to understand the toxicokinetics of plasma lead distribution for various exposure scenarios; that requires that acceptable methods be developed.

Bone lead measurement is the best way to assess body lead accumulation in such populations as high-risk urbanized young children, and it signals to the analyst and the policy-maker how rapidly lead is accumulating in bone. The role of this measure is enhanced considerably when techniques for integrating lifetime exposure are used in tandem with serial measurement of blood lead concentration in variably exposed young children or pregnant women.

Lead is released from bone in response to reductions in exposure or to metabolic stresses and alterations in the skeletal system. Lead appears in several compartments in bone, and these vary in their ability to release lead to the bloodstream. Furthermore, the known distribution characteristics of lead in bone compartments indicate that one can probe both long-term and shorter-term lead storage rates in bone spectrometrically. Such probing can be used both with populations in systematic development (children) and with populations in senescence (postmenopausal women).

Some biologic markers of effects of large exposures to lead have been characterized qualitatively and quantitatively and have been validated for toxicity in human populations. The usefulness of sensitive indicators, such as inhibition of particular enzymes, at very low lead concentrations has the added advantage that genetic susceptibility in lead-exposed subjects can be monitored. This would be the case for at least ALA-D activity distributions in sensitive populations exposed to lead. The sensitivity of some potentially useful biochemical markers has yet to be determined. Effect markers have been extremely helpful in identifying mechanisms of toxic action and in developing a better understanding of the biochemical interactions of lead in humans and other living systems.

Finally, little attention has been given to the identification of biologic markers of susceptibility. Current limited information does show that some children and other sensitive people could be predisposed to increased lead intoxication because of genetic disorders and nutritional factors. This is a subject of increasing importance to toxicologists and those responsible for public health.

5
Methods for Assessing Exposure to Lead

INTRODUCTION

The purpose of this chapter is to discuss analytic methods to assess exposure to lead in sensitive populations. The toxic effects of lead are primarily biochemical, but rapidly expanding chemical research databases indicate that lead has adverse effects on multiple organ systems especially in infants and children. The early evidence of exposure, expressed by the age of 6-12 months, shows up in prenatal or postnatal blood as lead concentrations that are common in the general population and that until recently were not considered detrimental to human health (Bellinger et al., 1987,1991a; Dietrich et al., 1987a; McMichael et al., 1988). As public-health concerns are expressed about low-dose exposures (Bellinger et al., 1991a,1987; Dietrich et al., 1987a; McMichael et al., 1988; Landrigan, 1989; Rosen et al., 1989; Mahaffey, 1992), the uses of currently applicable methods of quantitative assessment and development of newer methods will generate more precise dosimetric information on small exposures of members of sensitive populations.

Ultraclean techniques have repeatedly shown that previously reported concentrations of lead can be erroneously high by a factor of several hundred (Patterson and Settle, 1976). The flawed nature of some reported lead data was initially documented in oceanographic research: reported concentrations of lead in seawater have decreased by a factor of 1,000 because of improvements in the reduction and control of lead contamination during sampling, storage, and analysis (Bruland, 1983). Parallel decreases have recently been noted in reports on lead concentrations in fresh water (Sturgeon and Berman, 1987; Flegal and Coale, 1989).

Similar decreases in concentrations of lead in biologic materials have been reported by laboratories that have adopted trace-metal clean techniques. The decreases have been smaller, because lead concentrations in biologic matrixes are substantially larger than concentrations in water, and the amounts of contaminant lead introduced during sampling, storage, and analysis are similar. Nevertheless, one study revealed that lead concentrations in some canned tuna were 10,000 times those in fresh fish, and that the difference had been overlooked for decades because all previous analyses of lead concentrations in fish were erroneously high (Settle and Patterson, 1980). Another study demonstrated that lead concentrations in human blood plasma were much lower than reported (Everson and Patterson, 1980). A third demonstrated, with trace-metal clean techniques, that natural lead concentrations in human calcareous tissues of ancient Peruvians were approximately one five-hundredth those in contemporary adults in North America (Ericson et al., 1991).

Problems of lead contamination are pronounced because of the ubiquity of lead, but they are not limited to that one element. Iyengar (1989) recently reported that it is not uncommon to come across order-of-magnitude errors in scientific data on concentrations of various elements in biologic specimens. The errors were attributed to failure to obtain valid samples for analysis and use of inappropriate analytic methods. The former includes presampling factors and contamination during sampling, sample preservation and preparation, and specimen storage. The latter includes errors in choice of methods and in the establishment of limits of detection and quantitation, calibration, and intercalibration (Taylor, 1987).

Decreases in blood lead concentrations reportedly are associated with the decrease in atmospheric emissions of gasoline lead aerosols. The correlation between the decreases in blood lead and gasoline lead emissions is consistent with other recent observations of decreases in environmental lead concentrations associated with decreases in atmospheric emissions of industrial lead (Trefry et al., 1985; Boyle et al., 1986; Shen and Boyle, 1988). However, the accuracy of the blood lead analyses has not been substantiated by rigorous, concurrent intercalibration with definitive methods that incorporate trace-metal clean tech-

METHODS FOR ASSESSING EXPOSURE TO LEAD

niques in ultraclean laboratories. Moreover, previous blood lead measurements cannot be corroborated now, because no aliquots of samples have been properly archived. Nonetheless, within the context of internally consistent and carefully operated chemical research laboratories, valuable blood analyses have been obtained.

Future decreases in blood lead concentrations will be even more difficult to document, because the problems of lead contamination will be greater. Figure 5-1 depicts the relative amounts of blood lead and contaminant lead measured in people with high (50 μg/dL) and low (1 μg/dL) blood lead concentrations when amounts of contaminant lead introduced during sampling, storage, and analysis were kept constant (1 μg/dL each). The blood lead concentration measured in the person with a high blood lead concentration (53 μg/dL) will be relatively accurate to within 6%, because the sum of contaminant lead is small relative to blood lead. The same amount of contaminant lead, however, will erroneously increase the measured blood lead concentration of the other person by a factor of 4 (i.e., to 400%). That would seriously bias studies of lead metabolism and toxicity in the latter person. It would also lead to the erroneous conclusion that there was only about a 12-fold difference, rather than a 50-fold difference, in the blood lead concentrations of the two people. Both problems will become more important as the average lead concentration in the population decreases and as more studies focus on the threshold of lead toxicity.

In general, techniques to measure internal doses of lead involve measurement of lead in biologic fluids. Tissue concentrations of lead also provide direct information on the degree of lead exposure after lead leaves the circulation by traversing the plasma compartment and gaining access to soft and hard tissues. Once lead leaves the circulation and enters critical organs, toxic biochemical effects become expressed. It is of great importance for the protection of public health from lead toxicity to be able to discern the quantities of lead in target organs that are prerequisites for biochemically expressed toxic effects to become evident. The latter has been difficult, if not impossible, in humans, but lead measurement of the skeletons and placenta might make it more approachable with respect to fetuses, infants, women of child-bearing age, and pregnant women. Furthermore, measurements of lead in the skeleton of workers in lead industries has substantial potential for revealing the body burden of lead required for evidence of biochemical

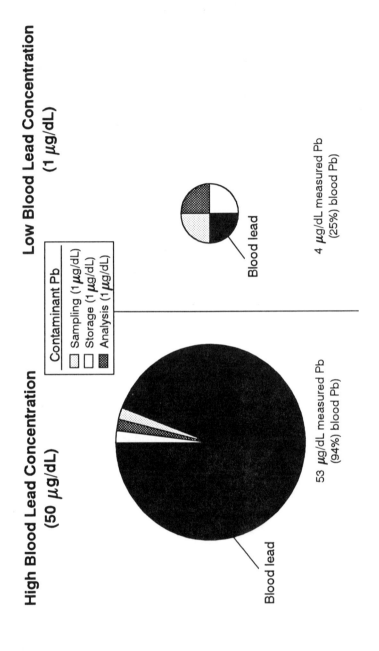

FIGURE 5-1 Illustration of relative problems of contamination in analysis of high and low blood lead concentration. Source: Adapted from Flegal and Smith, 1992.

toxicity to become manifest. Hence, measurements of lead in bone and placenta have the potential to couple quantitative analyses of lead at the tissue level to biochemical expressions of toxicity at the cellular level.

Noninvasive x-ray fluorescence (XRF) methods of measuring lead in bone, where most of the body burden of lead accumulates, have great promise for relating dosimetric assessments of lead to early biochemical expressions of toxicity in sensitive populations if their sensitivity can be improved by at least a factor of 10. The L-line XRF technique (LXRF) appears to be of potential value for epidemiologic and clinical research related to infants, children, and women of child-bearing age, including studies during pregnancy (Rosen et al., 1989, 1991; Wielopolski et al., 1989; Kalef-Ezra et al., 1990; Slatkin et al., 1992; Rosen and Markowitz, 1993). The K-line XRF method (KXRF) appears to be suited for studies in industrial workers and postmenopausal women, in addition to probing epidemiologic links between skeletal lead stores and both renal disease and hypertension (Somervaille et al., 1985, 1986, 1988; Armstrong et al., 1992).

Measurements of bone lead and blood lead in pregnant women throughout the course of pregnancy and assessments of amniotic-fluid lead concentrations and placental lead concentrations at term collectively hold promise for further characterizing the dynamics of maternal-fetal lead transfer.

Clinical research studies are examining epidemiologic issues related to the best measures of exposure and of the duration of exposure. Needleman et al. (1990) have reported that tooth lead concentrations constitute the best short- and long-term predictors of lead-induced perturbations in neurobehavioral outcomes. Longitudinal studies are examining whether cumulative measures (indexed by bone lead content based on LXRF), exposures during the preceding 30-45 days (indexed by blood lead concentrations), or exposures during critical periods are most important in the CNS effects of lead and in the reversibility of toxic effects on CNS function in children treated promptly with a chelating agent.

Some health effects of lead most likely depend on recent exposure; but knowledge of whether exposure was in the preceding few days, few months, or few years is extremely relevant clinically and epidemiologically. Previous reliance on blood lead concentrations alone has limited the use of time in treatment and outcome protocols. The half-time of lead in blood is short and reflects primarily recent exposure and absorption (Rabinowitz et al., 1976, 1977). Moreover, blood lead concentration does not reflect lead concentrations in target tissues that have different lead uptake and distribution or changes in tissue lead that occur when lead exposure is modified. Even lead in trabecular bone has a shorter duration than does lead in cortical bone. The most appropriate measure will likely vary with the end point in question. It is apparent, however, that current methods can strengthen epidemiologic and treatment efficacy studies by using multiple markers with different averaging times. The recent development of the ability to measure lead averaged over short periods (blood lead), intermediate periods (trabecular bone), and long exposure intervals (cortical bone) promises new techniques for measuring lead exposure in sensitive populations.

SAMPLING AND SAMPLE HANDLING

It is universally accepted that a crucial part of monitoring of lead in biologic material is the quality of sample collection and sample handling. Lead is pervasive and can contaminate samples randomly or systematically. In addition, the lead content of substances can be reduced by inappropriate collection, storage, or pretreatment. Protocols for acceptable sampling and sample handling vary with the material being sampled and the analytic technique being used, but most precautions apply across the board.

In all cases, sample containers, including their caps, must be either scrupulously acid-washed or certified as lead-free. That is particularly important for capillary- and venous-blood sampling as now incorporated into the guidelines of the 1991 CDC statement (CDC, 1991). For example, as little as 0.01 μg (10 ng) of contaminant lead entering 100 μL of blood in a capillary tube adds the equivalent of a concentration of 10 μg/dL, the CDC action level. Reagents added to a biologic sample before, during, or after collection especially must be lead-free. Lead concentrations in plasma or serum are generally so low to begin with and relative environmental lead concentrations so high that it is extremely difficult to collect and handle such samples without contamination. Urine sampling, especially the 8-hour or 24-hour sampling associated with chelator administration, requires collection in acid-washed containers. Although the amounts of lead being removed to urine with chelation are relatively high, the large volumes of sample and correspondingly large surface areas of collection bottles affect contamination potential.

A particularly important step in sample collection is the rigorous cleaning of the puncture site for capillary- or venous-blood collecting. The cleaning sequence of particular usefulness for finger puncture, the first step in blood lead screening, is that recommended by the Work Group on Laboratory Methods for Biological Samples of Association of State and Territorial Public Health Laboratory Directors (ASTPHLD, 1991). Fingers are first cleaned with an alcohol swab, then scrubbed with soap and water and swabbed with dilute nitric acid; and a silicone or similar barrier is used.

Sample storage is very important. Whole blood can be stored frozen for long periods. At -20°C in a freezer, blood samples can be stored for up to a year and perhaps longer.

Sample handling within the laboratory entails as much risk of contamination as sample collection in the field. Laboratories should be as nearly lead-free as possible. Although it is probably impractical for most routine laboratories to meet ultraclean-facility requirements (see, e.g., Patterson and Settle, 1976; EPA, 1986a), minimal steps are required, including dust control and use of high-efficiency particle-accumulator (HEPA) filters for incoming air and ultrapure-reagent use.

Collection and analysis of shed children's teeth entail unavoidable surface contamination, but this complication can be reduced by confining analysis to the interior matrix of a tooth, preferably the secondary (circumpulpal) dentin segment. The contaminated surface material is discarded. Isolation of the secondary dentin requires use of lead-free surfaces of cutting tools, lead-free work surfaces, and so forth.

MEASUREMENT OF LEAD IN SPECIFIC TISSUES

Whole Blood

The most commonly used technique to measure blood lead concentrations involves analysis of venous blood after chemical degradation (for example, wet ashing with nitric acid), electrothermal excitation (in a graphite furnace), and then measurement with atomic-absorption spectroscopy, or AAS (EPA, 1986a). With AAS, ionic lead is first vaporized and converted to the atomic state; that is followed by resonance absorption from a hollow cathode lamp. After monochromatic separation and photomultiplier enhancement of the signal, lead concentration is measured electronically (Slavin, 1988). Because it is much more sensitive than flame methods, the electrothermal or graphite-furnace technique permits use of small sample volumes, 5-20 µL. Physicochemical and spectral interferences are severe with flameless methods, so careful background correction is required (Stoeppler et al., 1978). Diffusion of sample into the graphite furnace can be avoided by using pyrolytically coated graphite tubes and a diluted sample applied in larger volumes.

Electrochemical techniques are also widely used for measurement of lead. Differential pulse polarography (DPP) and anodic stripping voltammetry (ASV) offer measurement sensitivity sufficient for lead analyses at blood concentrations characteristic of the average populace. The sensitivity of DPP is close to borderline for this case, so ASV has become the method of choice. It involves bulk consumption of the sample and thus has excellent sensitivity, given a large sample volume (Jagner, 1982; Osteryoung, 1988). This property is, however, of little practical significance, because, of course, sample size and reagent blanks are finite.

That ASV is a two-step process is advantageous. In the first step, lead is deposited on a mercury thin-film electrode simply by setting the electrode at a potential sufficient to cause lead reduction. The lead is thus concentrated into the mercury film for a specified period, which can be extended when higher sensitivity is needed; few techniques offer such preconcentration as an integral part of the process. After electrodeposition, the lead is reoxidized (stripped) from the mercury film by anodically sweeping the potential. Typically, a pulsed or stepping operation is used, so differential measurements of the peak current for lead are possible (Osteryoung, 1988; Slavin, 1988).

The detection limit for lead in blood with ASV is approximately 1 picogram (pg) and is comparable with that attainable with graphite-furnace AAS methods. The relative precision of both methods over a wide concentration range is $\pm 5\%$ (95% confidence limits) (Osteryoung, 1988; Slavin, 1988). As noted, AAS requires attention to spectral interferences to achieve such performance. For ASV, the use of human blood for standards, the presence of coreducible metals and their effects on the measurement, the presence of reagents that complex lead and thereby alter its reduction potential, quality control of electrodes, and reagent purity must all be considered (Roda et al., 1988). It must be noted, however, that the electrodeposition step of ASV is widely used and effective for reagent purification. The practice of adding an excess of other high-purity metals to samples, thereby displacing lead from complexing agents and ameliorating their concomitant interference effects, has demonstrated merit. Copper concentrations, which might be increased during pregnancy or in other physiologic states, and chelating agents can cause positive interferences in lead measurements (Roda et al., 1988).

The general sensitivity of ASV for lead has led to its use in blood lead analyses. The relative simplicity and low cost of the equipment has made ASV one of the more effective approaches to lead analysis.

As described in Chapter 4, the measurement of erythrocyte protoporphyrin (EP) in whole blood is not a sensitive screening method for identifying lead-poisoned people at blood lead concentrations below 50 μg/dL, according to analyses of results of the NHANES II general population survey (Mahaffey and Annest, 1986). Data from Chicago's screening program for high-risk children recently analyzed by CDC and the Chicago Department of Health indicated further the current limitations of EP for screening. The data, presented in Table 5-1, provide specificity and sensitivity values of EP screening at different blood lead concentrations. The sensitivity of a test is defined as its ability to detect a condition when it is present. The EP test has a sensitivity of 0.351, or about 35%, in detecting blood lead concentrations of 15 μg/dL or greater. This means that on average the EP test result will be high in about 35% of children with blood lead concentrations of 15 μg/dL or greater. It will fail to detect about 65% of those children. As the blood lead concentration of concern increases, the EP test becomes more sensitive. At blood lead concentrations of 30 μg/dL or greater, the sensitivity of the EP test is approximately 0.87. However, if it is used to detect blood lead concentration of 10 μg/dL or greater, the EP test has a sensitivity of only about 0.25.

The specificity of a test is defined as its ability to detect the absence of a condition when that condition is absent. As seen in Table 5-1,

TABLE 5-1 Chicago Lead-Screening Data, 1988-1989[a]

Definition of Increased Blood Lead, µg/dL	Sensitivity (Confidence Interval[b])	Specificity	Predictive Value Positive	Prevalence of Increased Blood Lead as Defined at Left
≥10	0.252 (0.211-0.294)	0.822	0.734	0.660
≥15	0.351 (0.286-0.417)	0.833	0.503	0.325
≥20	0.479 (0.379-0.579)	0.818	0.322	0.152
≥25	0.700 (0.573-0.827)	0.814	0.245	0.079
≥30	0.871 (0.753-0.989)	0.806	0.189	0.043
≥35	1.00 (0.805-)	0.794	0.119	0.030
≥40	1.00 (0.735-)	0.788	0.084	0.019
≥45	1.00 (0.590-)	0.782	0.049	0.011

| ≥50 | 1.00 | 0.775 | 0.014 | 0.003 |
| | (0.158-) | | | |

[a]Data indicate sensitivity, specificity, and predictive value positive of zinc protoporphyrin (ZPP) measurement for detecting increased blood lead concentrations. Increased ZPP is defined as ≥35 µg/dL. Definition of increased blood concentration varies. Data derived from systematic sample (2% of total) of test results for children 6 mo to 6 yr old tested in Chicago screening clinics from July 22, 1988, to September 1, 1989; these clinics routinely measure ZPP and blood lead in all children. n = 642. Data from M.D. McElvaine, Centers for Disease Control, and H.G. Orbach, City of Chicago Department of Health, unpublished; and McElvaine et al., 1991.

[b]Confidence intervals calculated by normal approximation to binomial method at 95% level for two tails. For estimates of sensitivity of 1.00, only lower-tail confidence interval is calculated. Exact binomial method is used.

roughly 83% of children with blood lead below 15 µg/dL will have a low EP result, and about 17% will have a high EP result. The test has a specificity of 0.83. The specificity of the test decreases as the cutoff increases. Because EP also increases in iron deficiency, a condition not uncommon among young children and occasional among pregnant women, the specificity of the EP test is reduced.

Although the sensitivity and specificity values appear higher than those obtained in the NHANES II population survey, in large part because of concurrent iron deficiency, the data confirm that unacceptably high numbers of children with increased blood lead concentrations will be missed by EP screening, particularly at blood lead concentrations below 25 µg/dL. Unfortunately, there is no feasible substitute for this heretofore convenient, practical, and effective tool as a primary screen. Measurements of alternative heme metabolites are available, but they require more extensive laboratory analyses and are largely surrogates of measurements of blood lead concentration.

Marked advances in instrumentation for blood-lead analysis during the last 10 years have yielded excellent precision and accuracy. The final limiting factors are now related specifically to technical expertise and cleanliness.

Plasma

Because of the high concentration of erythrocyte-bound lead, precautions must be taken to obtain nonhemolyzed blood when blood samples are collected for measuring the low concentrations of lead in plasma. Furthermore, an ultraclean laboratory setting and ultraclean doubly distilled column-prepared reagents are absolutely necessary (Patterson and Settle, 1976). Everson and Patterson (1980), using isotope-dilution mass spectrometry (IDMS) and strictly controlled collection and preparation techniques in an ultraclean laboratory, measured plasma lead concentration in a control subject and a lead-exposed worker. The control concentration of lead in plasma was 0.002 µg/dL, and that in the exposed worker was 0.20 µg/dL. Not surprisingly, these values were much lower than those obtained with graphite-furnace AAS; it can be concluded that graphite-furnace AAS methods do not yield sufficiently precise quantitative results for these measurements. Moreover, higher plasma lead concentrations have been reported even with IDMS (Rabinowitz et al., 1976); these results can be ascribed to problems in laboratory contamination. Collectively, therefore, it appears unlikely that measurement of lead in plasma can be applied widely to delineating lead exposure in sensitive populations.

Urine

The spontaneous excretion of lead in children and adults is not a reliable marker of lead exposure, being affected by kidney function, circadian variation, and high interindividual variation at low doses (Mushak, 1992). At relatively high doses, there is a curvilinear upward relationship between urinary lead and intake measures. Urinary lead is measured mainly in connection with the lead excretion that follow provocative chelation, i.e., the lead-mobilization test or chelation therapy in lead poisoning of children or workers. Such measurements are described later in this chapter.

Teeth

As noted elsewhere, shed teeth of children reflect cumulative lead exposure from around birth to the time of shedding. Various types of tooth analysis can be done, including analysis of whole teeth, crowns, and specific isolated regions.

Sampling and tooth-type selection criteria are particularly important. Teeth with substantial caries—i.e., over about 20-30% of surface area—should be discarded. Teeth of the same type should be selected, preferably from the same jaw sites of subjects in an epidemiologic study, to control for intertype variation (e.g., Needleman et al., 1979; Grandjean et al., 1984; Fergusson and Purchase, 1987; Delves et al., 1982). Replicate analyses are required, and concordance criteria are useful as a quality-assurance quality-control measure for discarding discordant values.

Preference in tooth analysis appears to lie with circumpulpal dentin, where concentrations are high. The higher concentrations in circumpulpal dentin are of added utility where effects are subtle. The use of whole teeth, crowns, or primary dentin is discouraged on two counts: random contamination is a problem on the surface of the outer (primary) enamel, and areas other than circumpulpal dentin are much lower in lead. Such problems would tend to enhance the tendency toward a null result in dose-effect relationships, i.e., Type II errors (Mushak, 1992).

In the laboratory, special care must be taken to avoid surface contamination, once sagittal sections of circumpulpal dentin have been isolated. A rinse with EDTA solution is helpful. For analysis, tooth segments are either dry-ashed or wet-ashed with nitric acid, perchloric acid, or a mixture of the two that is ultrapure as to lead contamination (e.g., Needleman et al., 1979). For calibration data, one can use bone powders of certified lead content (Keating et al., 1987).

Both AAS (Skerfving, 1988) and ASV (Fergusson and Purchase, 1987) methods have been used for tooth analysis. With AAS, both flame and flameless variations are often used. Lead concentrations are often high enough to permit dilution or chelation-extraction, thereby also minimizing calcium-phosphorus effects. Either method appears to be satisfactory (assuming that ASV entails complete acid dissolution of the tooth sample).

Although the use of tooth or dentin lead to assess the cumulative body burden of lead was an important discovery, it has practical limitations. It is necessary to wait for teeth to be shed and to rely on children to save them. Moreover, teeth are shed when children are 5-8 years old, so shed teeth cannot be used to estimate the body burden of lead in younger children. Furthermore, the newer capability to assess skeletal lead longitudinally with XRF improves the utility of tooth lead measurements.

Milk

To monitor daily lead intake from milk for metabolic balance studies or to estimate intake in areas where there is excessive external exposure to lead from other sources, milk samples should be collected in acid-washed polyethylene containers and frozen until analysis. After reaching room temperature, milk samples are sonicated and acid digested in a microwave oven, the residue is dissolved in perchloric acid, and the samples are subjected to AAS or ASV (Rabinowitz et al., 1985b).

Placenta

Placental lead measurements have been carried out after blotting of tissue and later digestion in acid at 110°C overnight. The dry residue was then dissolved in nitric acid, and lead is preferably measured with graphite-furnace AAS (Korpela et al., 1986). For placental and amniotic fluid measurements, standard reference materials are needed to ensure quality control in the laboratory. Furthermore, to assess placental lead concentrations accurately, it should be noted that region-specific concentrations of lead might exist in the placenta, and care must be taken to remove trapped blood in the placenta before analyses.

MASS SPECTROMETRY

The standard method by which all lead measurement techniques are evaluated is isotope-dilution thermal-ionization mass spectrometry (TIMS). Analyses of lead concentrations with this definitive method provide excellent sensitivity and detection limits. Recent analyses with this technique, when coupled with ultraclean procedures, have repeatedly demonstrated that many previously reported lead concentrations in biologic materials are erroneously high by orders of magnitude.

Mass spectrometers can also be used to measure stable lead isotopic compositions; to identify different sources of contaminant lead, from cellular to global; and to investigate lead metabolism without exposing people to radioactivity or artificially increased lead concentrations.

Mass spectrometers have a special niche in lead analyses, even though the measurements are relatively expensive, sophisticated, and time-consuming. Applications of mass spectrometers in analyses of lead in the biosphere are on the verge of being substantially broadened as a result of recent developments. Conventional TIMS is becoming much more sensitive and efficient. At the same time, inductively coupled plasma mass spectrometry (ICPMS) is becoming a relatively inexpensive and efficient alternative to TIMS and other established methods for elemental-lead analysis, and advances in secondary-ion mass spectrometry (SIMS), glow-discharge mass spectrometry (GDMS), and laser-microprobe mass analysis (LAMMA) have improved their capabilities for surface analysis and microanalysis of lead concentrations.

The primary differences among those techniques are the form of sample analyzed and the mechanism of introducing sample ions to a flight tube, where different isotopes are separated by magnetic and electric fields. In TIMS, a sample extract is deposited on a filament and then thermal ions are released by increasing the filament temperature within a vacuum. Sample solutions are ionized at atmospheric pressure with a highly ionized gas in ICPMS. Atoms in a solid source in an electric field are sputtered by an ionized gas and then thermalized by atom collisions in GDMS. Gas ions are also used as primary ions in SIMS, where they are focused on a solid surface to produce secondary ions by bombardment. Similarly, lasers are focused on a solid surface to vaporize, excite, and ionize atoms in microscopic areas of solid surfaces in LAMMA.

This section addresses both existing and projected applications of mass spectrometry in analysis of lead concentrations and isotopic compositions in biologic and environmental matrices. Definitive measurement of lead concentration with isotope-dilution TIMS is first reviewed. The use of stable lead isotopes to identify sources of contaminant lead and as metabolic tracers are then summarized. Finally, lead-related uses of four rapidly evolving types of mass spectrometry (ICPMS, SIMS, GDMS, and LAMMA) are briefly described.

Isotope-Dilution Mass Spectrometry

Lead concentrations in biologic tissues can only be approximated with analytic techniques, because no known analytic method can measure a true elemental concentration in any matrix. The definitive methods, including isotope-dilution mass spectrometry (IDMS), use TIMS. Relatively accurate measurements are derived with reference methods, which are calibrated with standard reference materials. Less-accurate measurements are acquired with routine methods, which provide assigned relative values to judge the analyzed results. The hierarchy in accuracy of these different analytic techniques is listed in Table 5-2.

The most accurate method of analyzing lead concentrations in biologic matrices is IDMS, which is independent of yield and extremely sensitive and precise (Webster, 1960). Mass spectrometric analyses distinguish lead from false signals by measuring the relative abundances

TABLE 5-2 Hierarchy of Analytic Methods with Respect to Accuracy

Analytic Data	Analytic Method
True value	No known method
Definitive value	Definitive method, e.g., IDMS
Reference-method value	Reference method
Assigned value	Routine method

Source: Heumann, 1988. Reprinted with permission from *Inorganic Mass Spectrometry*; copyright 1988, John Wiley & Sons.

of the four stable lead isotopes (^{204}Pb, ^{206}Pb, ^{207}Pb, and ^{208}Pb). False signals in lead-isotope measurements are identified by simultaneous measurements of fragment ions in the adjacent masses (203 and 205 amu). IDMS is superior to any reference method that requires separate measurements of lead signal intensities for sample and reference materials. The advantages and primary analytic disadvantages of IDMS are summarized in Table 5-3. In IDMS, a spiked sample enriched in one isotope (^{206}Pb) is prepared. The ^{206}Pb-to-^{208}Pb isotopic ratio of the spiked sample is then measured in a mass spectrometer. The ratio of ^{206}Pb to ^{208}Pb is then used to determine the concentration of lead in the original sample.

IDMS analyses, like all other elemental analyses, require a correction for the contaminant-lead blank. This includes all contaminant lead added during sampling, storage, and analyses (Patterson and Settle, 1976). The individual contribution of each of those contaminant-lead additions to the total lead signal for each analysis must be determined separately for highly accurate measurements. Blank measurements are especially appropriate for IDMS, because high concentrations of sensitivity and precision are required to measure the lead concentrations of "trace-metal-clean" reagents and containers accurately. That is illustrated in Table 5-4, a tabulation of lead blank measurements for blood lead analyses in a trace-metal-clean laboratory.

Lead concentration measurements with IDMS must also correct for

TABLE 5-3 Advantages and Disadvantages of Lead Concentration Analyses with IDMS

Advantages	Disadvantages
Offers precise and accurate analysis	Is destructive
Permits nonquantitative isolation of the substance to be analyzed	Requires chemical preparation of sample
Offers ideal internal standardization	Is time-consuming
Multielement, as well as oligoelement and monoelement, analyses possible	Is relatively expensive
Offers high sensitivity with low detection limits	

Source: Heumann, 1988. Reprinted with permission from *Inorganic Mass Spectrometry*; copyright 1988, John Wiley & Sons.

isotopic variations of lead, including natural variations among samples and isotopic fractionation during the analyses. The latter correction, common to all elemental analyses with IDMS, is addressed with standard IDMS techniques (Heumann, 1988). The former, which is relatively unusual among the heavier elements, requires separate isotopic analyses of unspiked samples. That necessitates additional analyses for lead concentration measurements, but it also provides unique applications of lead isotopic composition measurements that are addressed in the following section.

Lead Isotopic Composition in the Identification of Lead Sources

Measurable differences in stable lead isotopic compositions throughout the environment are caused by the differential radioactive decay of ^{238}U ($t_{1/2}$ = 4.5 x 10^9 years), ^{235}U ($t_{1/2}$ = 0.70 x 10^9 years), and ^{232}Th

TABLE 5-4 Quantification of Lead Contamination in Analyses of Blood Lead Concentrations with Isotope-Dilution Thermal-Ionization Mass Spectrometry and Trace-Metal-Clean Techniques[a]

Procedure	Reagent and Container	Volume, mL	Lead Concentration, pg/mL	Lead Blank, pg
Sample digestion	HNO_3	3	4	12
	$HClO_4$	1	10	10
	Teflon bomb	-	-	5
	Total digestion blank			27
Microcolumn extraction	HBr	4.8	7	34
	HCl	2.4	5	12
	HNO_3	0.2	4	1
	$HClO_4$	0.1	10	1
	Teflon microcolumns	-	-	3
	Teflon vials	-	-	5
	Total extraction blank			56
Mass-spectrometer loading blank				1
Total treatment and analytic blank				84

[a] Data from Flegal and Smith, 1992.

($t_{1/2}$ = 1.4 x 10^{10} years) to form ^{206}Pb, ^{207}Pb, and ^{208}Pb, respectively (Faure, 1986). The fourth stable lead isotope, ^{204}Pb, has no long-lived radioactive parent. Stable lead isotopic compositions differ among geologic formations with different ages, parent-daughter isotope ratios, and weathering processes.

There is no measurable biologic, chemical, or physical fractionation of lead isotopes in the environment, and natural differences in lead isotopic compositions in geologic formations persist after the lead has been extracted and processed (Russell and Farquhar, 1960; Barnes et al., 1978). Differences in the lead isotopic composition of urban aerosols in the United States, for example, are shown in Table 5-5. The differences reflect regional and temporal variations in isotopic compositions of ores used for lead alkyl additives in the United States, emissions of other industrial lead aerosols in the United States, and the atmospheric transport of foreign industrial lead aerosols to the United States (Patterson and Settle, 1987; Sturges and Barrie, 1987).

Because there is no measurable isotopic fractionation of lead in the biosphere, sources of industrial lead can be identified by their isotopic composition (Flegal and Stukas, 1987). For example, Yaffee et al. (1983) used stable lead isotopic compositions to identify the primary source of lead (soils contaminated with leaded paint) in a group of lead-poisoned children in Oakland, California. Other investigators have also used the technique, which was pioneered by C.C. Patterson, to identify sources of contaminant lead in U.S. populations (Table 5-6).

Stable Lead Isotopic Tracers in Metabolic Studies

Most tracer studies of lead metabolism have used radioisotopes (^{203}Pb, ^{210}Pb, and ^{212}Pb) of lead (Table 5-7). Those studies have been few, because the half-lives of radioisotopes of lead (e.g., ^{203}Pb, 51.88 hours; ^{210}Pb, 22.5 years; and ^{212}Pb, 10.64 hours) are not generally conducive to this type of research. Moreover, applications of any radioisotopes in metabolic studies are now severely limited: even minimal radiation exposure is avoided unless it is clinically necessary.

The analytic and clinical limitations of using lead radioisotopes as metabolic tracers, in conjunction with recent advances in the analytic

TABLE 5-5 ^{206}Pb:^{207}Pb Ratios of Aerosols in the United States

Location	Year	^{206}Pb:^{207}Pb[a]	Reference
Houston, Tex.	1970	1.220	Chow et al., 1975
St. Louis, Mo. (urban)	1970	1.230	Rabinowitz and Wetherill, 1972
St. Louis, Mo.	1970	1.220	Rabinowitz and Wetherill, 1972
Berkeley, Calif.	1970	1.199	Rabinowitz and Wetherill, 1972
Benecia, Calif.	1970	1.157	Rabinowitz and Wetherill, 1972
San Diego, Calif.	1974	1.211	Chow et al., 1975
Narragansett, R.I.	1986	1.196	Sturges and Barrie, 1987
Boston, Mass.	Pre-1981	1.191	Rabinowitz, 1987
Boston, Mass.	1981	1.207	Rabinowitz, 1987
West Coast	1963	1.143	Shirahata et al., 1980
West Coast	1965	1.153	Chow and Johnstone, 1965
West Coast	1974	1.190	Patterson and Settle, 1987
West Coast	1978	1.222	Shirahata et al., 1980
Midwest	1982-1984	1.213	Sturges and Barrie, 1987
Midwest	1986	1.221	Sturges and Barrie, 1987

[a]95% confidence limit of ^{206}Pb:^{207}Pb measurements ≤0.005.

capabilities of mass spectrometry, have given new impetus to the use of stable lead isotopes in this type of analysis. The primary advantage of using stable isotopes as tracers is that neither subjects nor researchers are exposed to radiation. Different sources of exposure and metabolic processes can be monitored simultaneously, because there are three independent isotopes of stable lead for which ratios can be calculated. Because only very small amounts of stable isotopic tracers are required for highly precise analyses, the potential for metabolic perturbations due to large exposures to lead is minimized.

The high precision and accuracy required for lead isotopic tracer studies have made TIMS the method of choice. That has been recognized in several recent reviews of the use of stable isotopes in metabolic

TABLE 5-6 Studies Using Lead-Isotopic Compositions as Tracers of Environmental and Biologically Accumulated Contaminant Lead in United States

Lead Sources and Organisms	Reference
Industrial aerosols	Chow and Johnstone, 1965 Patterson and Settle, 1987 Sturges and Barrie, 1987
Surface waters	Flegal et al., 1989
Coastal sediments	Ng and Patterson, 1981, 1982
Terrestrial ecosystems	Shirahata et al., 1980 Elias et al., 1982
Lead-smelter emissions and equines	Rabinowitz and Wetherill, 1972
Seawater and marine organisms	Smith et al., 1990 Flegal et al., 1987
Paint lead and children	Yaffee et al., 1983 Rabinowitz, 1987
Aerosols and humans	Manton, 1985, 1977

studies, even though TIMS analyses are slower than most other methods and require rigorous pretreatment to eliminate organic contaminants and interfering elements (Janghorbani, 1984; Turnlund, 1984; Hachey et al., 1987; Janghorbani and Ting, 1989a). Newer types of mass spectrometry (e.g., ICPMS) have not yet achieved the sensitivity and precision of TIMS, which are required for most analyses of lead isotopic compositions in biologic and environmental matrices. Other types of mass spectrometry still in development (e.g., resonance-ionization mass spectrometry) might become applicable to lead isotopic tracer studies. Conversely, gas-chromatography mass spectrometry (GC-MS) does not appear to be a likely option, because its maximal attainable precision is insufficient for reliable isotope-ratio determinations.

TABLE 5-7 Lead-Metabolism Studies Using Radioisotopes

Radio-isotope	Study Subject	Reference
^{203}Pb	Bone cell lead-calcium interactions	Rosen and Pounds, 1989
^{203}Pb	Gastrointestinal lead absorption	Watson et al., 1986
^{203}Pb	Lead retention	Campbell et al., 1984
^{203}Pb	Gastrointestinal lead absorption	Blake and Mann, 1983
^{203}Pb	Gastrointestinal lead absorption	Blake et al., 1983
^{203}Pb	Lead absorption	Flanagan et al., 1982
^{203}Pb	Gastrointestinal lead absorption	Heard and Chamberlain, 1982
^{203}Pb	Oral lead absorption	Watson et al., 1980
^{203}Pb	Gastrointestinal lead absorption	Blake, 1976
^{210}Pb	Osteoblastic lead toxicity	Long et al., 1990
^{212}Pb	Gastrointestinal lead absorption	Hursh and Suomela, 1968

Inductively Coupled Plasma Mass Spectrometry

One of the major advances in analyses of both lead concentrations and isotopic compositions has been the recent development of inductively coupled plasma mass spectrometry (ICPMS). It has rapidly assumed a prominent position in many research laboratories since the first commercial instrument was introduced in 1983 (Houk and Thompson, 1988). There has been a nearly exponential increase in ICPMS publications in the last two decades (Hieftje and Vickers, 1989), and it has recently been identified as one of the "hottest areas" in science (Koppenaal, 1990).

Although ICPMS is recognized as potentially the most sensitive multielemental method, it is still not established. It has not yet permitted widespread and inexpensive analyses of lead in laboratories with routine procedures—especially in medical research, where applications of ICPMS are still in their infancy—and it has yet to be used in any manner other than cursory and illustrative examples (Janghorbani and Ting, 1989a).

The advantages and limitations of ICPMS are discussed in several recent reviews (e.g., Houk and Thompson, 1988; Kawaguchi, 1988; Koppenaal, 1988, 1990; Marshall, 1988). A typical multielement analysis with ICPMS takes only a few minutes to conduct. In addition, the spectra are relatively simple, because there are only a few different molecular ions and doubly-charged ions produced in this technique, compared with other multielement techniques (Russ, 1989). However, the precision and sensitivity of ICPMS are still poorer than those of TIMS. ICPMS has not demonstrated the capacity to produce high-precision ($\pm 1\%$) measurements of lead-isotope ratios in biologic matrices (Ward et al., 1987; Russ, 1989). Other limitations include problems of internal and external calibration, interference effects, deposition of solids during sample introduction, requirements for microgram quantities of the element of interest, and loss in electron multiplier amplification.

Calibration problems—which have been detailed by Puchelt and Noeltner (1988), Vandecasteele et al. (1988), Doherty (1989), and Ketterer et al. (1989)—are being resolved on several fronts (Koppenaal, 1988, 1990). Beauchemin et al. (1988a,b,c) have demonstrated the applicability of external standards for parts-per-billion analyses of biologic materials. Additionally, numerous investigators are now addressing the need to compare ICPMS with other analytic techniques (Hieftje and Vickers, 1989), including the first intercalibrated measurements of lead isotopic compositions in blood (Delves and Campbell, 1988; Campbell and Delves, 1989).

Other investigators are studying the fundamentals of ICPMS and the applicability of different plasmas, nebulizers, and techniques to minimize problems caused by dissolved solids in samples with inherently high solid contents (Koppenaal, 1988, 1990). The latter techniques include flow-injection (Dean et al., 1988; Hutton and Eaton, 1988), complexation-preconcentration (Plantz et al., 1989), and ion-exchange (Lyons et al., 1988) techniques. They could be adapted to investigate different isotopes of lead in biologic fluids (i.e., blood and urine).

The potential of ICPMS for research in medical research has been recognized in recent reviews (e.g., Dalgarno et al., 1988; Delves, 1988). There have already been several reports of relatively good agreement between intercalibrated measurements of blood lead concentrations based on ICPMS and other established analytic techniques. For example, Douglas et al. (1983), and Brown and Pickford (1985) found good agreement in results of analyses with ICPMS and graphite-furnace AAS. Diver et al. (1988) have investigated the applicability of albumin as a reference material in an intercalibration of ICPMS, AES and AAS. There have also been numerous intercalibrations of elemental concentrations in different biologic matrices based on ICPMS and other techniques (e.g., Douglas and Houk, 1985; Munro et al., 1986; Pickford and Brown, 1986; Ward et al., 1987; Beauchemin et al., 1988a,b,c; Berman et al., 1989).

The most extensive study of ICPMS measurements of lead isotopic compositions in medical research has investigated selenium metabolism (Janghorbani et al., 1988; Janghorbani and Ting, 1989b). Dean et al. (1987) measured lead isotopic compositions in milk and wine, and Caplun et al. (1984), Date and Cheung (1987), Longerich et al. (1987), Sturgis and Barrie (1987), and Delves and Campbell (1988) measured lead isotopic compositions in other environmental matrices. However, most ICPMS measurements of lead isotopic compositions have not included ratios for the least common lead isotope, ^{204}Pb (1.4% of the total mass), and that is required for definitive isotopic composition analysis. It is also generally recognized that ICPMS measurements are not precise enough to show significant variations in most lead isotopic compositions (Ward et al., 1987; Russ, 1989).

In spite of the limitations just described, the phenomenal advances in ICPMS instrumentation in the last 2 decades indicate that it has passed through a remarkably short adolescence and is now in a mature stage with a prolific record of instrument acquisition and application (Koppenaal, 1990). Widespread application of ICPMS in medical research and public-health studies of lead contamination and metabolism might still require improvement, including the development of alternative sample introduction techniques, such as vapor generation, recirculating nebulizers, ultrasonic nebulizers, and electrochemical furnaces. The sensitivity of ICPMS analyses must also be improved to the point where nanogram quantities of lead are sufficient to determine isotopic ratios to the limits of the instrument's precision (0.1%), so that counting statistics will not limit the precision of the analysis. That will require faster and more linear ion detectors, such as the Daly detector (Huang et al.,

1987), and higher-resolution mass analyzers, such as more sophisticated quadrupoles or magnetic sector instruments, to obtain the resolution needed to separate polyatomic and oxide peaks from elemental isotopes (Gray, 1989). Corresponding reductions in noise and turbulence in the plasma might also involve the use of other gases for plasma support (Montasser et al., 1987; Satzger et al., 1987). Other improvement is needed in multiplier longevity and analytic precision (Russ, 1989).

Secondary-Ion Mass Spectrometry

Secondary-ion mass spectrometry (SIMS) has recently been recognized as a major technique for surface composition analyses and microstructural characterization (Lodding, 1988), because of its high sensitivity and good topographic resolution, both in depth and laterally. The applicability of SIMS for analysis of lead in environmental and biologic matrices has not been fully realized, because too few instruments are available for that type of analysis and their accuracy is too low. This was discussed in a recent review of atomic mass spectrometry by Koppenaal (1988), who noted that SIMS was popular in electronics and the materials sciences, but not in environmental and biologic fields. His comprehensive review of articles on SIMS analysis in those two fields was limited to seven references, and only one of those involved lead analysis in organisms (Chassard-Bouchaud, 1987). Koppenaal (1990) later indicated that some analyses with SIMS might be replaced with glow-discharge mass spectrometry (GDMS), because SIMS had a "notoriously dismal reputation" for accuracy. The potential for SIMS analysis of lead in biologic and environmental samples remains in question.

Glow-Discharge Mass Spectrometry

Advances in glow-discharge mass spectrometry have indicated its potential for elemental analyses in solid matrices. That potential has not been realized, because of the high cost of GDMS instrumentation, difficulties of direct solids analyses, and low accuracy of current GDMS measurements. However, those obstacles have recently been reduced by the introduction of relatively inexpensive quadrupole-based GDMS and a pronounced increase in the number of studies on the applicability of GDMS analyses in a variety of matrices. The evolving applicability of GDMS analysis is summarized in recent reviews (Sanderson et al., 1987, 1988; Harrison, 1988; Harrison and Bentz, 1988; Koppenaal, 1988, 1990). GDMS might soon succeed spark-source mass spectrometry and SIMS for bulk-solids analysis. GDMS analyses are highly precise (SD $\pm 5\%$) in the low parts-per-billion range. Although the accuracy of GDMS (± 10-300%) is still too variable for analysis of lead in environmental samples (Huneke, 1988; Sanderson et al., 1988), quantitative results ($\pm 5\%$) appear possible (Harrison, 1988). Therefore, GDMS might soon become a valuable technique for monitoring elemental concentrations in geologic matrices, including lead in contaminated soils.

Laser-Microprobe Mass Spectrometry

Applications of lasers in medical research continue to expand, as reported in two reviews by Andersson-Engels et al. (1989, 1990). Others have provided complementary reviews of advances in laser-microprobe mass spectrometry and related laser-microprobe techniques (e.g., Koppenaal, 1988, 1990; Verbueken et al., 1988). Those methods use lasers as an alternative to electron and ion beams for localized chemical analysis. They have been incorporated in laser-microprobe mass analysis (LAMMA), laser-induced mass analysis (LIMA), laser-probe mass spectrography (LPMS), scanning-laser mass spectrometry (SLMS), direct-imaging-laser mass analysis (DILMA), time-resolved laser-induced breakdown spectroscopy (LIBS), laser-ablation and laser-selective excitation spectroscopy (TABLASER) and laser-ablation and resonance-ionization spectrometry (LARIS).

Laser techniques have several advantages for microprobe analyses of lead concentrations in biologic matrices, including its high detection efficiency (about 10^{-20} g), speed of operation, spatial resolution (1 μm), capabilities for inorganic and organic mass spectrography, and potential for separate analysis of the surface layer and core of particles. Disadvantages of the technique are that it is destructive, the quality of the light-microscopic observation is poor, and quantification of elemental concentrations is questionable. Some of those features are listed in Table 5-8 in comparison with other microanalytic techniques, including electron-probe x-ray microanalysis, secondary-ion mass spectrometry, and Raman microprobe analysis.

Linton et al. (1985) have reviewed laser- and ion-microprobe sensitivities for the detection of lead in biologic matrices. Their analyses indicate that the relative detection limit of lead in biologic material with a lateral resolution of 1 μm with LAMMA (5 μg/g) is about 100 times better than that obtained with a Cameca IMS-3F ion microscope and that the useful yield of lead ions was about 100 times better with LAMMA (10^{-3}) than with the ion microscope (10^{-5}). The sensitivity for lead in LAMMA is also much better than in SIMS, when normalized to potassium (Verbueken et al., 1988).

LAMMA was developed specifically to complement other microanalytic techniques for determining intracellular distributions of physiologic cations and toxic constituents in biologic tissues. It has already been used to investigate the distribution of lead in various tissues while still in a development stage in studies of the localization of lead in different cell types of bone marrow of a lead-poisoned person (Schmidt and Ilsemann, 1984), of the topochemical distribution of lead across human arterial walls in normal and sclerotic aortas (Schmidt, 1984; Schmidt and Ilsemann, 1984; Linton et al., 1985), and of the distribution of lead in placental tissue and fetal liver after acute maternal lead intoxication. The potential of LAMMA in medical and environmental research has been summarized by Verbueken et al. (1988), who concluded:

> We have every reason to expect that routine quantitative LAMMA analysis will develop reasonably successfully within the next few years. Important in the achievement of quantitative accurate analysis will be necessary fundamental studies of the processes involved in the measurement, including laser-matter interactions, plasma chemistry, and physics.

ATOMIC-ABSORPTION SPECTROMETRY

Atomic-absorption spectrometry (AAS) in routine use has been available to laboratories for almost 30 years, and it has been established in its current instrumented forms for about 20 years (Ottaway, 1983;

Van Loon, 1985; EPA, 1986a; Shuttler and Delves, 1986; Miller et al., 1987; Angerer and Schaller, 1988; Osteryoung, 1988; Slavin, 1988; Delves, 1991; Jacobson et al., 1991; Mushak, 1992). In brief, AAS involves thermal atomization of lead from some transformable sample matrix, absorption of radiation by the sample's lead-atom population (at one of lead's discrete wavelengths) from some element-specific source, and minimization or removal of diverse spectral interferents to provide a clean, lead-derived detection signal. The source has typically been a lead-specific hollow cathode or electrodeless discharge lamp.

AAS methods that use commercially available equipment are generally single-element methods. Simultaneous analyses of lead and other elements usually require different analytic approaches and are beyond the scope of this report. As is also the case with anodic-stripping voltammetry (described later), lead is routinely measured in biologic media as a concentration of the total element. Speciation methods have used AAS-based metal-specific detectors in which AAS units are interfaced with additional instruments, e.g., gas-liquid and liquid chromatographs (Van Loon, 1985; EPA, 1986a).

Much of the methodologic improvement in AAS in the last 30 years has centered on the nature of the thermal excitation and the relative efficiency in controlling spectral interference. The evolution of commercially available atomic-absorption (AA) spectrometers over the last 30 years has concerned principally improvements in atomization and detection. Originally, liquid samples containing the lead analyte were aspirated into the flame of an AA spectrometer. That approach was generally unsatisfactory for trace analysis because a large sample was required and the detection limit was too high. Often, preconcentration was required by such means as chelation and extraction with an organic solvent.

In 1970, the Delves Cup microflame technique appeared on the scene; it was a marked improvement with respect to detection-limit and sample-size requirements (Delves, 1977, 1984, 1991). The Delves Cup approach also helps to minimize the spectral interferents that result from an organic matrix if one uses a preignition step (Ediger and Coleman, 1972). As noted by Delves (1991), this approach does not lend itself to automation, but a high sample throughput is nonetheless achievable. Accuracy and precision are quite satisfactory. Although this technique

TABLE 5-8 Summary of Characteristics of Four Types of Microprobes[a]

Characteristic	X-Ray Microanalysis	Ion Microprobe (Ion Microscope)[b]	Laser Microprobe (Transmission Geometry)	Raman Microprobe
Probe	Electrons	Ions	Photons (laser)	Photons (laser)
Detection method	Characteristic X-rays: WDS or EDS	Ions ("+" or "-"); Double-focusing MS	Ions ("+" or "-"); TOF MS	Photons (Raman); Double monochromator; PM
Resolution of detection	WDS, 20 eV; EDS, 150 eV	$M/\Delta M = 200{,}10{,}000$	$M/\Delta M = 800$	$0.7\ cm^{-1}$ (spectr); $8\ cm^{-1}$ (image)
Lateral resolution (analyzed area)	WDS, 1 μm; EDS, 500–1,000 Å[c]	1–400 μm	1 μm	1 μm
Imaging (spatial resolution)	SEM, 70 Å; STEM, 15 Å	0.5 μm (SII)	1 μm	1 μm
Information depth	≤ 1 μm	Tens of Å	--	--
Detection limits	WDS, 100 ppm; EDS, 1,000 ppm	ppm[d]	ppm	Major comp.
Elemental coverage	WDS, Z ≥ 4; EDS, Z ≥ 11	H–U	H–U	--

Isotope detection	No	Yes	No
Compound information	No	Yes[e]	Yes
In-depth analysis	No	Yes	No
Destructive	No	Yes	No
Quantitative analysis	Yes	(Yes)	Yes

Additional columns: Yes / Yes / Difficult / Yes / (Yes)

[a]EDS: energy-dispersive spectrometer.
PM: photomultiplier.
SEM: scanning electron microscope.
SII: secondary-ion image.
STEM: scanning transmission electron microscope.
TOF-MS: time-of-flight mass spectrometer.
WDS: wavelength-dispersive spectrometer.
[b]Ion microscope Cameca-3f.
[c]High-concentration deposit.
[d]Depends on element of interest and on chemical environment, including nature of primary ion beam (Ar^+, O^+_2, O^-, Ca^+).
[e]More in "static SIMS."
Source: Verbueken et al., 1988. Reprinted with permission from *Inorganic Mass Spectrometry*; copyright 1988, John Wiley & Sons.

is not as popular as flameless GF-AAS, it performs well in competent hands.

Over the last 10-15 years, GF-AAS has become the most favored analytic variant of AAS. Liquid-sample requirements are modest (10 μL routinely), and sensitivity is 10 times that of the Delves Cup method or better and up to 1,000 times that of conventional aspiration-flame analysis. Across various sample types, detection is readily achievable below the parts-per-billion level; in simple matrices, the detection limit is around 0.05-0.5 ppb. High sensitivity permits matrix modification with diluents. GF-AAS has the added advantage of adaptability to automation, and the literature contains a number of examples of such adaptation for analyzing large numbers of samples (e.g., Van Loon, 1985).

A potentially persistent problem with all micro-AAS methods is spectral interference of diverse physicochemical and matrix types, and the phenomenon might be most problematic with GF-AAS. Various means of interference control, termed *background correction*, have been developed. The deuterium-arc correction was first, and removal of spectrometric interference with a Zeeman-effect spectrometer was second; other approaches are being developed (Van Loon, 1985).

AAS has a number of general advantages for the clinical-chemistry and analytic-toxicology laboratory with respect to quantitative measurements of lead in biologic media. Equipment in various forms is generally available commercially at moderate cost. Methods are relatively straightforward and impose only moderate requirements for expert personnel. AAS, particularly in the form of GF-AAS but also in the Delves Cup flame AAS (micro-FAAS) variation, has the requisite analytic sensitivity and specificity for whole blood, provided that overall laboratory proficiency is appropriate. Of particular importance, AAS has a well-established track record of accuracy and precision and has been adopted for reference-laboratory use in proficiency testing and external quality-assurance programs. Therefore, likely sources of problems for an analyst using AAS will probably have been identified and described elsewhere. AAS has performed well in ordinary laboratories, according to results of external proficiency-testing and quality-assurance programs.

ANODIC-STRIPPING VOLTAMMETRY AND OTHER ELECTROCHEMICAL METHODS

The electrochemical technique of anodic-stripping voltammetry (ASV) in its trace analytic applications has been available for about 20 years and is based on early studies of polarography by Heyrovsky and colleagues in the 1920s (see, e.g., Nurnberg, 1983). Its current analytic characteristics for lead in various media, developed from the studies of Matson (Matson et al., 1971; Nurnberg, 1983; Stoeppler, 1983a; EPA, 1986a; Delves, 1991). Unlike AAS, ASV can be used as a multielement quantitation technique, provided that deposition and stripping characteristics are favorable.

ASV works on a two-step principle of lead analysis. First, lead ion liberated from some matrix is deposited by a two-electron reduction on a carbon-supported mercury film as a function of time and negative voltage. The deposited lead is then reoxidized and electronically measured via anodic sweeping. Those electrochemical processes generate current-potential curves (voltammograms) that can be related to concentration of the metal, e.g., lead ion. In common with all electrochemical techniques, ASV measures total quantity. Given infinite time and sample volume, electrochemical methods would theoretically have infinite sensitivity. In practice, both laboratory time and available sample sizes are limited, and this results in finite sensitivity. The ASV process collects all the metal of interest at the deposition step, which is the principal factor in ASV's high operational sensitivity (low detection limit).

ASV, like most electrochemical techniques, is affected by the thermodynamic and oxidation-reduction characteristics of the analytic matrix. In practice, the lead ion must be liberated into a chemical matrix that permits deposition and stripping without interference. Early methods used wet chemical degradation of organic matrices, including biologic media, such as whole blood (Matson et al., 1971; Nurnberg, 1983; EPA, 1986a). That approach, although it ensured mineralization, introduced a high risk of contamination by lead and was time-consuming. More recently, decomplexation has involved use of a mixture of competitive ions, including chromium and mercury, which liberate lead from binding sites by competitive binding. This approach poses problems of adequate liberation; it gives low readings at concentrations below 40 µg/dL and high readings at high concentrations (Delves, 1991). In the case of blood analyses, however, one can effectively overcome the problems by extending the time of decomplexing and carefully resolving lead and interfering copper signals (Roda et al., 1988).

Two methodologic advantages of ASV make it popular in laboratories that measure lead in various media, including whole blood. The advantages parallel those of GF-AAS and explain why ASV and GF-AAS are the two most widely used techniques for measuring lead in biologic media in the United States and elsewhere. First, the necessary accuracy and precision required for trace and ultratrace analysis of heavy metals are achievable with ASV in competent hands, according to many reports and a performance record. Second, the sensitivity is such that it is appropriate for the low average lead concentrations in media now being encountered. As in the case of AAS, detection limits are medium-specific; in simple media, such as water, 10-100 pg can be measured. Required analyst expertise is modest, and the equipment is commercially available at costs lower than those of AAS.

ASV is not the only electrochemical method that can be used for quantitative analysis of lead. Jagner et al. (1981) reported lead measurement in blood with computerized potentiometric-stripping analysis (PSA); the reported sensitivity with the plating time used was reported to be 0.5 µg/dL, and within-run precision was reported to be 5% at a mean blood lead of 7 µg/dL. Another technique, differential-pulse polarography (DPP), has not found a wide reception, although its analytic characteristics and advantages approach those of ASV (Angerer and Schaller, 1988). New instrumentation uses this method, but its long-term utility for routine analysis of biologic media needs to be determined.

NUCLEAR MAGNETIC RESONANCE SPECTROSCOPY

Nuclear magnetic resonance spectroscopy (NMR) is finding increased application in the study of lead, for three specific reasons. First, the development of high-magnetic field instruments has enhanced the sensitivity of NMR measurement of lead in biologic systems. Second, the evolution of specialized techniques for resolving complex spectra has made it possible to obtain clear spectra on solid, as well as liquid, samples. Finally, the increasing availability of NMR instruments for computed axial tomographic (CAT) scanning has opened up new potential for diagnostic exploration involving lead deposition.

The use of NMR for lead characterization has been aptly reviewed by Wrackmeyer and Horchler (1989). The NMR properties of lead are determined by the presence of ^{207}Pb (abundance, 22.6%). Studies have indicated that resonance frequencies of ^{207}Pb are influenced by the molecular environment around that isotope. Thus, the observed chemical shifts range over several thousand parts per million, as opposed to the few parts per million observed for ^{1}H or ^{13}C. With this broad spectral range, identification of specific chemical forms of lead is simplified because of the increased resolution. One can anticipate increased application of NMR in the study of lead speciation in both environmental and clinical settings. The ultimate potential of the technique in these settings is yet to be defined; it will be limited by the expense of the equipment and the high level of operating expertise required.

Some preliminary conclusions concerning the basic mechanisms of early toxicity at the cellular level have been reached with ^{19}F NMR. ^{19}F NMR on biologic samples containing the reagent 1,2-bis(2-amino-5-fluorophenoxy)ethane-N,N,N',N'-tetraacetic acid, initially described by Smith et al. (1983), has been developed to measure intracellular free calcium and free lead simultaneously in the rat osteoblastic osteosarcoma cell line ROS 17/2.8 (Schanne et al., 1989, 1990a,b). Treatment of the cells with lead produced marked increases in intracellular free calcium; and concurrent measurements of intracellular free lead yielded a concentration of about 25 pM (Schanne et al., 1989). This in vitro technique in clinical studies is useful in the measurement of intracellular free calcium and lead in erythrocytes of exposed persons. It can also help in characterizing the dose-response relationships in intracellular free calcium transport.

In addition to its ability to characterize molecular mechanisms of lead toxicity at the cellular level in humans, ^{19}F NMR can also be used for simultaneous measurement of free cytosolic lead, calcium, zinc, iron, and other metals. Even if such a discrete and early marker of lead toxicity at the cellular concentration is uncovered, the high expense of instrumentation and the technical expertise required for NMR will limit its application to selective studies of women, children, and adult workers in lead industries.

THE CALCIUM-DISODIUM EDTA PROVOCATION TEST

The calcium-disodium EDTA ($CaNa_2EDTA$) provocation test is a diagnostic and therapeutic test to ascertain which children with blood lead concentrations between 25 and 55 µg/dL will respond to the chelating agent $CaNa_2EDTA$ with a brisk lead diuresis (Piomelli et al., 1984; CDC, 1985). Children with a brisk response qualify for a 5-day hospital course of $CaNa_2EDTA$ (Markowitz and Rosen, 1984; Piomelli et al., 1984; CDC, 1985). The test constitutes a chemical biopsy of exchangeable lead, which is considered to be the most toxic fraction of total body lead (Chisolm et al., 1975, 1976). $CaNa_2EDTA$ is confined to the extracellular fluid, and lead that is excreted originated primarily in bone and, to a lesser extent, in soft tissues (Osterloh and Becker, 1986).

An 8-hour provocation test in an outpatient department has been shown to be as reliable as a 24-hour test (Markowitz and Rosen, 1984); and measurement of urinary lead excretion with graphite-furnace AAS (for example) is convenient and accurate, if urine is collected quantitatively in a lead-free apparatus. However, the provocation test is impractical, cumbersome, and labor-intensive in young children, from whom quantitative urine collection is very difficult. EDTA chelation has potentially toxic side effects in children and the lead redistribution caused by chelation might actually increase the toxicity of lead in some target tissues, such as brain. Therefore, it is unlikely that the test will ever find wide application in sensitive populations.

In a recent study that used blood lead concentration and net corrected LXRF photon counts as predictors of $CaNa_2EDTA$-test outcomes, 90% of lead-poisoned children were correctly classified as $CaNa_2EDTA$-positive or -negative with a high degree of specificity and sensitivity (Rosen et al., 1989). Therefore, LXRF with blood-lead concentration measurement might ultimately replace the $CaNa_2EDTA$ test and prove suitable for assessing large numbers of individual subjects in sensitive populations to select those at risk for toxic effects of lead. Furthermore, new effective oral chelating agents might substantially advance the treatment of childhood lead intoxication. However, it is likely to be difficult to use such agents for outpatient treatment, unless lead-poisoned children are treated in a clean environment (transition housing) without excessive lead ingestion, so that increasing lead absorption by the use of an oral chelating agent will be precluded.

X-RAY FLUORESCENCE MEASUREMENT

During the last 2 decades, methods have been developed to measure lead in bone noninvasively. The residence time of lead in bone is long, and these methods could broaden the range of information available in biologic monitoring of lead exposure to reflect long-term body stores associated with chronic exposure and thereby complement plasma and whole-blood lead measurements, which respond principally to acute exposure.

Bone lead measurements might or might not be related directly to adverse biologic effects of lead exposure. As with any emergent technology, their utility needs to be assessed both with respect to unanswered questions about lead effects and with respect to other means of predicting them. The physiologic availability of lead from different bone compartments has received little investigation.

The relation between bone lead measurements and exposure or biologic dose of lead is not clear. There is some evidence that cumulative exposure can be estimated from bone lead measurements; but what this means, both for lead monitoring and for research into health effects of lead exposure, has yet to be explored in full detail.

The techniques for in vivo measurement of bone lead with XRF can be divided into two groups, and there are variations within each group. The major difference between the two groups is that one relies on detecting K-shell x rays and the other on L-shell x rays for lead measurement. Radiation from an x-ray machine or other radiation source penetrates to the innermost shells of lead atoms, thereby ejecting either an L- or K-shell electron. As a result of the filling of a vacancy left by an ejected electron, a K- or L-fluorescent x ray is emitted; and the emitted energy is characteristic of the element that absorbed the original x ray, in this case lead. A detection system that collects, counts, displays, and analyzes emitted x rays according to their energy is thus able to determine how much lead is present in the sample. The characteristics of L-line and K-line XRF techniques are summarized in Table 5-9. Both techniques use γ rays or x rays of low energy. Low energy is below about 150 keV. In this energy range, a photon can undergo three types of interaction: photoelectric, Compton, and elastic (or coherent).

In photoelectric interaction, the incoming photon gives up all its energy to an inner-shell electron of an atom. The electron is ejected from the atom with a kinetic energy equal to the energy of the incoming photon minus the electron binding energy. The result is a vacancy in the inner electron shell of the atom. The vacancy is filled by a less tightly bound electron, and energy is released either as an x ray or by the ejection of valence electrons that have kinetic energy. Detecting the energy of the x ray is the basis for the quantitative measurement of lead in bone.

In Compton scattering, the incoming photon interacts with an electron (usually a valence, loosely bound, electron), and the photon energy, minus a small amount of electron binding energy, is then divided between the electron in the atom (as kinetic energy) and the resulting emitted photons. The energy of the electron and the scattered photon depends on the angle of scatter, and the scattered photons are the most intense feature of XRF spectra. There are two strategies for minimizing the extent to which Compton-scattered photons limit the precision of XRF. One uses the angular dependence of Compton scattering and a choice of energy of incoming photons to minimize the interference of this scattering in the measurement of lead x rays (Ahlgren et al., 1976; Somervaille et al., 1985). The other produces a polarized incoming x-ray beam (Wielopolski et al., 1989). Such x rays have the property that they cannot be Compton-scattered at right angles to their plane of polarization; thus their interference in the measurement of lead x rays is minimized.

In elastic scattering, the incoming photon interacts with an atom as a whole. There is a change in direction, which is usually slight, and there is virtually no energy absorption, so the photon continues with, in effect, its full energy. The probability of elastic scattering depends on photon energy, angle of scatter, and atomic number of the scattering atom. It is more probable for low-energy photons, for small scattering angles, and for high atomic numbers. In XRF spectra, it produces photons nearly equal in energy to the incoming photons and interferences that are smaller than those produced by Compton scattering but larger than those from lead x rays from samples with lead concentrations in the range of biologic interest. The importance of elastic scattering for XRF lies in the fact that its strong dependence on atomic number means that it is provoked primarily by calcium and phosphorus, and not by lead, in biologic tissue samples containing bone. This feature of elastic scattering can be used to standardize lead concentration to bone in practical measurements (Somervaille et al., 1985).

Dosimetry

Bone lead XRF measurements require irradiation with ionizing radiation, so the radiation dose and associated risk are important. The dose depends on the amount of energy absorbed and the mass of tissue in which it is absorbed. In this case, photoelectric and Compton interactions contribute to dose. The dose also depends on which tissues absorb the energy; different tissues have different sensitivities to ionizing radiation. For fluorescence measurements, one could consider the maximum skin dose, the total energy deposited in tissue, or the effective dose (ICRP, 1991). In measuring effective dose, full account is taken of the different radiosensitivities of tissues. Both LXRF and KXRF have recently been subjected to detailed dosimetric analysis, and the results show that the two techniques give comparable doses for measurements of young children, but that adult doses are lower with both techniques (Kalef-Ezra et al., 1990; Slatkin et al., 1991,1992; Todd et al., 1992). Effective doses in children and adults reported for the KXRF method were less than those obtained with LXRF (Todd et al., 1992; Slatkin et al., 1992). The equivalent dose to a conceptus from one KRF measurement, 38-52 nSv, is about 20 times greater than that from an LXRF measurement (Slatkin et al., 1992). These dosimetric inconsistencies between pregnant women and children have not been addressed for the KXRF method (Todd et al., 1992). Although dosimetric assessments of KXRF and LXRF instruments followed ICRP 60

TABLE 5-9 Characteristics of L-Line and K-Line XRF Instruments[a]

Characteristic	L-Line	K-Line
Fluorescing source	Low-energy generator: incident photons are polarized[b]	^{109}Cd[c]
Energy of imparted x rays	20 keV[b]	88 keV[c]
Dosimetry, μSv		
Infants	2.5[d,e]	1.1[d,f]
Children	1.0[b,d]	0.4[d,f]
Teenagers	0.5[d,e]	0.2[d,f]
Adults	0.3[d,e]	0.04[d,f]
Dosimetry during pregnancy	-0.003% of natural background radiation during 9-mo pregnancy[g]	-0.002% of natural background radiation during 9-mo pregnancy[f]
Minimal detection limit with 3 mm of overlying skin	4 ppm[h]	6 ppm[i]
Validation with whole limbs	Yes	Yes
Type of bone sampled	Cortex[h]	Cortex and trabecular[i]
Counting time	16.5 min[b]	30 min[c,j]
Counts corrected for bone mineral content	No	Yes
Replicate reproducibility in vivo after repositioning of instrument	±2 ppm[k] (95% confidence limits)	Not yet published

TABLE 5-9 (cont.)

Characteristic	L-Line	K-Line
Anticipated improvements	Modify polarizer and detector[l] to decrease counting time to <5 min and increase sensitivity	Use larger-volume hyperpure germanium detectors and faster electronics to increase count rate
	Express data as Pb:Sr ratios	Modify geometry to narrow Comptom peak, reducing background

[a]Although dosimetric assessments of KXRF and LXRF instruments followed ICRP 60 guidelines (ICRP, 1991), these estimates were obtained through different protocols, conditions and assumptions. Accordingly, reported equivalent doses for complementary techniques are not absolutely comparable.
[b]Wielopolski et al., 1989; Rosen et al., 1989, 1991, 1993; Rosen and Markowitz, 1993.
[c]Somervaille et al., 1985, 1986.
[d]Dosimetry calculated according to ICRP guidelines (ICRP, 1991).
[e]Slatkin et al., 1991, 1992.
[f]Todd et al., 1992.
[g]Kalef-Ezra et al., 1990.
[h]Wielopolski et al., 1989.
[i]Somervaille et al., 1986.
[j]Somervaille et al., 1988.
[k]Slatkin et al., 1991.
[l]Wielopolski et al., 1989; Rosen et al., 1989.
[m]Green et al., 1993.

guidelines (ICRP, 1991), they used some necessarily different considerations and assumptions based on the particular instruments being assessed. Those differences might be less important than the magnitude of the doses, which are 1-5 μSv for children and as low as 50 nsv for adults measured with the K x-ray system. Annual natural background radiation doses are around 3 mSv, so at worst the dose of a small child is equivalent to less than 1 day's natural background radiation. Alternatively, the additional radiation can be equated to one-tenth that involved in a single transcontinental air flight, which arises from the increase in the cosmic-ray component of natural background with altitude. In terms of risk, the additional chance of a cancer death, the most probable of the damaging effects of low to moderate radiation exposure, is around 1 in 10 million, compared with a natural rate of about 1 in 5. In this context, it seems clear that radiation exposure is not a limiting hazard in the use of either of these measurement techniques.

Volume Sampled

KXRF and LXRF techniques differ in the volume of material sampled. That is because attenuation of photons in tissue depends on the photon energy. The higher energies used in KXRF mean that photons penetrate farther into tissue. However, the greatest difference stems from the attenuation of the characteristic x rays emerging from the tissue. LXRF quantitation is based on the 10.55-keV L_α x rays; KXRF quantitation uses x rays of 72.8- 85 keV, with the most prominent being the $K_{\alpha 1}$ at 75.0 keV.

These photons do not have a definite range, so it is convenient to characterize their ability to penetrate by the thickness of material required to reduce their intensity by a given factor. If one uses a factor of 2 and compares the attenuation of only the L_α x rays and the combined effect of attenuation of incoming 88-keV photons and $K_{\alpha 1}$ x rays, the half-value thickness for L x rays is 1.6 mm in soft tissue (muscle) and 0.35 mm in bone, and the half-value thickness for K x rays is 19.0 mm in muscle and 9.0 mm in bone. That comparison might markedly overestimate the difference between the two techniques. Other physical factors are extremely important, such as the relative "hardness of" energy of the exciting radiation and especially the relative magnitude of signal-to-noise ratios, as determined by the magnitude of background counts under peaks. KXRF and LXRF systems have not been compared experimentally in that respect.

Precision

Quoted values for precision of the two techniques are similar and are in the range ±2-10 µg/g. In making comparisons, it should be noted that L x-ray precision and results are usually stated in terms of lead concentration in wet bone, whereas K x-ray results are usually stated as lead concentration in bone mineral. For adult tibia, that introduces a factor of about 1.8; that is, 5 µg of lead per gram of wet bone corresponds to 9 µg of lead per gram of bone mineral. For other bones, the factor is greater. The precision varies with lead location, soft tissue attenuation, count rates, and signal-to-noise ratios. Direct instrument comparisons could further assess the relative importance of these variables for the KXRF and LXRF techniques. Such comparisons between instruments of both types are increasingly relevant as instrument geometry is modified (Green et al., 1993). The quantitative limits of LXRF and KXRF instruments are approximately 2-5 and 10 µg/g, respectively. These instruments assess lead concentrations at different bone sites, inasmuch as they use different depths and areas of bone. It is unclear whether the areas of bone differ in the extent to which lead can be mobilized by physiologic processes, such as growth, or pathologic processes, such as osteoporosis.

The adequacy of quantitation limits of existing methods can be judged, in part, by comparison with bone lead concentrations of susceptible populations, e.g., young children and women of child-bearing age. The most extensive such data sets are those of Barry (1975), Drasch et al. (1987), and Drasch and Ott (1988).

Bone samples obtained at autopsy were analyzed with atomic-absorption spectrophotometry, in the case of the German data, and dithizone clorimetry, in the case of the Barry (1975) analyses. Barry (1975) reported that children living in the United Kingdom had mean lead concentrations of 4 µg/g wet weight (or approximately 8 µg/g ash weight). Drasch et al. (1987) and Drasch and Ott (1988) investigated bone lead concentrations in a nonoccupationally exposed population of adults and children living near Munich, Germany. In infants, the geometric mean lead concentration was less than 1 µg/g wet weight in temporal bone, femur, and pelvic bone. Bone lead concentrations in people 10-20 years old had a geometric mean lead concentration under 2 µg/g in temporal and pelvic bones and midfemur. In 240 nonoccupationally exposed adults, the predominantly cortical femur and temporal bone had geometric mean lead concentrations of about 3.9 and 5.6 µg/g, respectively. The predominantly trabecular pelvic bone had a geometric mean lead concentration of less than 2 µg/g. Comparison of those data from persons who came to autopsy during 1983-1985 with data obtained in 1974 from subjects who lived in the same geographic area showed that bone lead concentrations had decreased 3-5 times for children and approximately 1.3 times for adults.

On the basis of those data from the United Kingdom and Germany, nonoccupationally exposed subjects had bone lead concentrations under the limits of XRF instruments in use in the early 1990s. Both LXRF and KXRF might have little clinical applicability for children who have typical environmental lead exposures. Currently, neither technique can provide accurate data on groups other than highly exposed populations, e.g., workers in lead industries and other adults with prolonged high exposures. However, some investigators are using these techniques to examine those with relatively high environmental exposures. To measure bone lead concentration in the general population, XRF would need increased sensitivity, perhaps increased by a factor of 5 to 10.

Nonetheless, XRF does provide information on past lead exposures of a portion of the population. For example, XRF measurements would complement screening programs aimed at high-risk urban children or workers in lead industries. Such information would supplement identification of more highly exposed persons through clinical history or demographic methods. In addition, evidence of excessive exposure to source-specific lead in affected communities could be produced by comparative XRF surveys of heavily exposed and well-controlled, normally exposed cohorts. The question of the relationship of endogenous release of lead from bone to future health risks is important in a determination of where XRF methods are likely to be focused in the future. It is known that release of lead from bone in occupationally exposed adults not only increases blood lead, but increases markers of hematotoxicity caused by lead (Alessio et al., 1976a; Alessio, 1988). Bone lead concentrations are strongly linked to the diagnosis of lead poisoning and to the efficacy of treatment for poisoning in lead workers (Christoffersson et al., 1986). Children with known lead exposure who were immobilized during hospitalization showed a marked increase in blood lead to concentrations associated with poisoning (Markowitz and Weinberger, 1990).

Accuracy

The lead concentration will normally be measured by reference to calibration standards. XRF sensitivity varies with position in the sample, so assumptions have to be made about the spatial variation of lead concentration in the sample being measured. Adequate calibration standards can be produced for either XRF technique, and this should not be a major source of inaccuracy; but it should be noted that it might be difficult and expensive to produce a single set of artificial calibration standards that would serve for both LXRF and KXRF.

A typical assumption about lead distribution is that lead concentration in soft tissue is zero and that lead is uniformly distributed in bone. The first part of the assumption is probably reasonable, given the much higher concentration of lead in bone than in soft tissue. Some consideration might be given to the fact that soft tissue overlying tibia is more sensitively sampled than the bone itself; data on this subject are currently lacking.

Some data show that bone lead concentration is relatively uniform along a tibia (Wittmers et al., 1988), but other data suggest that some variations in concentration may occur with depth in the tibia (Lindh et al., 1978; Schidlovsky et al., 1990). Depth varies by around 0.1 mm or less, so the different spatial responses of the two systems can be significant, in that the half-value thickness for L x rays in bone is 0.35 mm, whereas that for K x rays is 9 mm. The K x-ray system would integrate over the spatial variations in lead concentration, thus rendering the measurement relatively unaffected by such discontinuities, but also obscuring any information that might be conveyed by the variations. Definitive studies concerning the possibility of differences in the microlocalization of lead in bone have not yet been published.

Practical XRF Systems

The first bone lead XRF measurements used 122- and 136-keV γ rays from ^{57}Co to excite the K x rays and measured lead in a finger bone (Ahlgren et al., 1976). A similar approach was taken, apparently independently, in measuring tooth lead in vivo (Bloch et al., 1977), and the finger-bone system was also adopted elsewhere (Price et al., 1984). The original ^{57}Co system has now been in use since 1971, and longitudinal data on retired lead workers cover most of that time (Nilsson et al., 1991).

A later development in KXRF was the use of γ rays from ^{109}Cd to excite the x rays. This had the advantage that the γ-ray energy, at 88 keV, was so close to the minimum energy required to excite lead K x rays that the normalization of lead to bone mineral with the elastically scattered photons was very accurate (Somervaille et al., 1985). The ^{109}Cd approach has now been adopted, with a number of variations, by laboratories in Wales (Morgan et al., 1990), Finland (Erkkilä et al., 1992), and the United States (Jones et al., 1987; Hu et al., 1990), as well as the original laboratory in England (Chettle et al., 1991).

LXRF systems have emerged from the collaboration of two U.S. laboratories. The original work used ^{109}Cd (and silver x rays, rather than the 88-keV γ ray) or ^{125}I (Wielopolski et al., 1983), but the same researchers later developed an improved version in which a better detection limit was achieved with polarized x rays to reduce the extent of Compton scattering observed in the region of the lead x rays (Wielopolski et al., 1989).

The polarized LXRF system and KXRF with ^{109}Cd have the best published performance, and they form the basis of most of the comments made here. However, it must be noted that further work is in progress on both systems, and improvements in both systems appear likely.

Validation

LXRF measurements have been validated with whole human limbs obtained after amputation. Measurements were recorded with the overlying tissue intact and then on the bare bone; samples of the bone were then sent for independent chemical analysis (Wielopolski et al., 1989). Such a protocol is in principle highly desirable because of its potential rigor. In this case, nine surgically amputated limbs were used. Clearly, additional assessments of limbs for measurements with KXRF and LXRF instruments would increase the systematic validation of both techniques.

KXRF validation has thus far been largely indirect. Bare-bone samples have been analyzed by subjecting small core samples to AAS and then sending the residue of the bone for KXRF analysis (Somervaille et al., 1986). Paired analyses have been performed in that way on 80 samples, incorporating metatarsals, calcaneus, tibia sections, and tibia fragments. The two analyses were independently calibrated. The mean difference between XRF and AAS results on all 80 samples was less than 0.1 μg of lead per gram of bone mineral, and the maximum difference observed in any subset of the data was about 5 μg/g for a group of three tibia sections (Chettle et al., 1991). The only flaw in the validation procedure was that it was indirect; extension to in vivo measurements relied on other experiments, which showed that the accuracy of the K x-ray normalization to elastic scatter was independent of overlying tissue thickness (Somervaille et al., 1985). In some cases, whole limbs were measured with KXRF and then analyzed with AAS; the whole-limb results (three samples) tend to confirm the accuracy of the technique (Hu et al., 1990). A few direct comparison data came from subjects who died within a few months of in vivo measurement (Christoffersson et al., 1984; Nilsson et al., 1991). They show good agreement between in vivo and autopsy data, but the data are few, time necessarily passed between in vivo and autopsy measurements, and, because the subjects were elderly, lead and calcium metabolism might have been changing during the last few months of life.

Proponents of both LXRF and KXRF systems are clearly satisfied with the validation of their measurements, but there is room for rigorous validation conducted under the auspices of an independent external laboratory, rather than based on the internal procedures of a single laboratory. These systems should be systematically assessed before their clinical utility for screening sensitive populations is evaluated, as described below.

- Photon spectra from XRF instruments should be carefully measured, including entrance dose, imparted energy, and effective dose (ICRP, 1991).
- Minimal detection limits (MDLs) should be established with standard reference materials or amputated limbs whose lead content

parallels that in a specific sensitive population that might be studied. MDLs and calibration procedures should be developed under standard instrument operating conditions to obtain clinically usable data.

- The effective dose of x rays in sieverts should be calculated by combining absorbed-dose data with geometric measurements on subjects of various age groups while they are being measured for bone lead under standard instrument operating conditions. The calculations must be expressed according to the most current National Council on Radiation Protection and Measurements (NCRP) and International Commission on Radiological Protection (ICRP) guidelines, with weighting factors for all relevant radiosensitive tissues.

- The precision of in vivo XRF measurements should be determined from duplicate measurements in up to 50 subjects in a sensitive population to determine the 95% confidence limits of each measurement. For replicate measurements, the tibia should be repositioned at a site distal or proximal to the original measurement before obtaining the second measurement. The tibia has been used most frequently in clinical measurements.

- Standard values for effective dose (ICRP, 1991) should be recalculated and compared by independent experts from data on LXRF and KXRF instruments operating under conditions of comparable in vivo precision and similar bone lead detection limits.

- After those systematic procedures have been appropriately completed, the clinical utility of XRF instrumentation should be carefully evaluated to ascertain whether clinically relevant data are being obtained.

- In all the procedures noted above, the general radiation—exposure principle of ALARA (as low as reasonably achievable) should be strictly adhered to.

Clinical Uses of X-Ray Fluorescence

On the basis of published data (Somervaille et al., 1985; Rosen et al., 1989, 1991, 1993; Wielopolski et al., 1989; Kalef-Ezra et al., 1990; Slatkin et al., 1991, 1992; Todd et al., 1992; Rosen and Markowitz, 1993), the committee concludes that the utility and efficacy of the KXRF and LXRF measurement techniques are similar and the radiation dose associated with both methods is relatively small for all ages.

As noted previously, lead has a residence time in blood of only 30-45 days (Rabinowitz et al., 1976, 1977). Hence, measurements of lead in blood reflect recent exposure. A blood lead concentration might be more useful when lead exposure can be reliably assumed to have been at a given concentration, as in occupationally exposed adults, than when intermittent exposure is occurring or has occurred, as in children exposed to leaded paint. With the recent development of *complementary* K-line and L-line XRF techniques to measure bone lead stores noninvasively (Somervaille et al., 1985, 1986, 1988; Rosen et al., 1989, 1991; Wielopolski et al., 1989; Rosen and Markowitz, 1993), it now appears possible to assess directly a time-averaged compartment of lead in bone, where its residence time is months to years; however, wide clinical application of XRF still needs to be developed (Rabinowitz et al., 1976, 1977; Chamberlain et al., 1978). XRF has the potential to relate bone lead stores (and blood lead concentration), as predictive outcome measures, to the development of early expressions of lead toxicity, such as biochemical, electrophysiologic, and neurobehavioral indexes (Rosen et al., 1989, 1991; Rosen and Markowitz, 1993). In a pioneering study, Needleman et al. (1979) found that short-term and long-term neurobehavioral deficits caused by lead were closely correlated to lead concentrations in compact bone, as reflected in tooth lead concentrations (Needleman et al., 1990). Moreover, because blood lead concentrations decrease once excessive lead exposure ends and a course of chelation treatment has been successfully completed, blood lead concentration is likely to underestimate high bone concentrations (Christoffersson et al., 1986; Rosen et al., 1991; Rosen and Markowitz, 1993).

KXRF methods developed in Sweden (Ahlgren et al., 1980) and England (Somervaille et al., 1985, 1986, 1988; Chettle et al., 1989; Armstrong et al., 1992) have demonstrated the clinical utility of bone lead measurements in industrially exposed adults. The minimal detection limit for lead, with 3 mm of overlying tissue, was about 10 μg/g of wet bone (Somervaille et al., 1988). Specifically, studies by the English group showed that KXRF measurements are a good indicator of long-term lead exposure, as assessed with the cumulative blood lead concentration index in 20 control subjects and 190 workers in lead industries (Somervaille et al., 1988). Moreover, Christoffersson et al. (1986), using the KXRF technique, showed progressive decreases in tibial lead content with a mean half-time (short-term and long-term) of about 7 years, once workers were removed from lead industries. Both skeletal storage compartments contributed to lead in blood even many years after the end of high occupational exposure (Christoffersson et al., 1986; Schütz et al., 1987a). With the KXRF technique in male workers in lead industries, it is possible to measure lead concurrently in cortical and trabecular bone (Chettle et al., 1989). The two sites yield different residence times for lead.

The clinical utility of the LXRF technique was first demonstrated in 59 lead-poisoned children (blood lead concentrations of 23-53 µg/dL) (Rosen et al., 1989, 1991). The minimal detection limit for lead, with 3 mm of overlying tissue, was about 4 µg/g of wet bone (Wielopolski et al., 1989). In the Rosen et al. work, 90% of "lead-poisoned" children were correctly classified as being $CaNa_2EDTA$-positive or $CaNa_2$-EDTA-negative, and the majority of the children had bone lead concentrations at least as high as those measured in normal and industrially exposed adults. The data indicate that bone lead concentrations in the study children were about 1,200-3,700 times greater than those in ancient Peruvian bones (Ericson et al., 1979). Hence, by the age of 7 years, mildly exposed children have already achieved enormous skeletal burdens of lead, which are likely to have profound effects on their health (Emmerson and Lecky, 1963; Patterson, 1980; Pirkle et al., 1985; Thompson et al., 1985; Bellinger et al., 1987,1988; Landis and Flegal, 1988; Silbergeld et al., 1988; Rosen et al., 1989, 1991; Markowitz and Weinberger, 1990; Rosen and Markowitz, 1993). The LXRF technique allowed decreases in bone lead content to be followed sequentially during chelation therapy of children and after correction of leaded-paint hazards.

Bone lead concentrations in the Rosen and Markowitz (in press) prospective study of 162 untreated children, some of whom qualified for $CaNa_2EDTA$ therapy (Piomelli et al., 1984; CDC, 1985), are being coupled to several other outcome measures, including biochemical, electrophysiologic, and neurobehavioral characteristics. Bone lead concentrations were measured with LXRF in children whose homes had undergone lead abatement; some of them had undergone successful chelation therapy, as judged by conventional criteria (including return of blood lead and erythrocyte protoporphyrin concentrations to acceptable concentrations). Their bone lead concentrations were 3-5 times higher than concentrations measured in apparently normal European children

(Rosen et al., 1991; Rosen and Markowitz, 1993). The clinical utility of LXRF has been extended to estimate bone lead content in lead-exposed and non-lead exposed suburban populations (Rosen et al., in press). The mean bone lead value in 269 residents of the highly exposed suburb (15 ppm) was 3-fold greater than that of the reference suburb (5 ppm). These differentially exposed populations included measurements in children, teenagers, and adults.

During pregnancy of lead-exposed women, bone lead stores could have an unfavorable effect on the neurologic maturation of fetuses or infants through physiologic demineralization of maternal bone after 12 weeks of pregnancy or during lactation. The potential intergenerational impact of lead might be present in women exposed to high concentrations of lead early in life.

The in vivo precision of LXRF measurements was determined from duplicate measurements in 37 randomly selected children who had tibial bone lead concentrations of 8-47 μg/g (mean, 16 $\pm$ 1 SEM). Each pair of measurements was performed on the same day with the tibia repositioned 4 cm distal to the site of the first measurement. The 95% confidence limits of replicate measurements were $\pm$2 μg/g (Slatkin et al., 1991). For use of the LXRF technique in children 1 and 5 years old, the effective doses were calculated to be 2.5 and 1.0 μSv, respectively—a radiation exposure much less than that for one dental x-ray picture (Wielopolski et al., 1989; Slatkin et al., 1991,1992; Rosen and Markowitz, 1993). Similar or lower values have been reported for the KXRF (Todd et al., 1992). Those effective doses were calculated according to the current NCRP and ICRP recommendations (NCRP, 1989; Rosen et al., 1989; Wielopolski et al., 1989; ICRP, 1991; Slatkin et al., 1991, 1992; Todd et al., 1992). Moreover, dosimetry data concerning the use of the L-line technique during pregnancy have shown that the radiation exposure of the conceptus is negligible—no more than 0.003% of the natural background ionizing radiation absorbed by the average American human conceptus during a full-term pregnancy (Kalef-Ezra et al., 1990).

As further improvements in the detector and polarizer are made, the minimal detection limit and photon counting time (less than 5 minutes) of the LXRF technique are expected to decrease severalfold without an increase in radiation exposure (Rosen et al., 1989). Until the decrease in counting time is achieved, children less than 4 years old will continue to be mildly sedated with orally administered chloral hydrate, so that they remain still during the 16.5-minute measurement. Chloral hydrate, which is used extensively in pediatric practice for mild sedation before many electrophysiologic procedures, is not known to have any substantial short-term side effects when appropriately administered. However, more recent concerns regarding its potential carcinogenicity have been raised by the findings of laboratory animal studies. The potential long-term effects have yet to be resolved. The usefulness of the LXRF technique can probably be expanded by measuring the ratio of the L-line bone lead concentration to the K-line signal in the 10- to 16-keV interval of the XRF spectrum (Rosen et al., 1989).

Standard reference materials are now needed for external and internal instrument calibrations for both the L-line and K-line techniques. The calibrations should be carried out under strictly defined operating conditions that achieve the minimal detection limit of each instrument with concurrent measurement of radiation exposure, according to recommendations of the ICRP (1991). Internal and external calibrations should be assessed directly by independent experts. Calibration, with the dosimetry and systematic measurements noted earlier, should provide confidence that risk assessments of both L-line and K-line techniques have been thorough.

L-line and K-line XRF techniques are *complementary* and provide a new, exciting, and needed capability to assess lead exposure that has accumulated over time (many months to several years) in sensitive populations. Both techniques are based on the general principles of x-ray fluorescence, but the current characteristics of each technique indicate that each has specific applications for developing needed information on different sensitive populations.

Given the current state of development of both the L-line and K-line methods, it is not currently recommended that either instrument be used as a screening technique in general populations.

Research Needs

Although L-line and K-line XRF methods are becoming standard techniques to assess previous lead exposure over a person's lifetime, they entail critical research needs that must be addressed before they can be more generally applied for screening of populations.

- Even though the radiation doses are low for both techniques, efforts should be made to develop methods to reduce radiation exposures further.
- Measurements of randomly selected men, in the near future, with both L-line and K-line instruments, will provide important information as to whether the two techniques estimate the same or different compartments of lead in bone. Men have been designated for initial comparisons, because dosimetry data on men have been detailed and published for the L-line technique, and estimates of dosimetry do not appear to be widely dissimilar for the K-line method in the same population.
- It is possible that the two methods measure lead from metabolically different skeletal compartments. Accordingly, experimental microlocalization studies of human limbs would be relevant. Such studies could be carried out with proton-induced x-ray emission (PIXE) after careful sectioning of limbs and assessment of tissues that are sectioned by experts and do not measure more than about 30-40 μm.
- Before the K-line XRF method can be considered for use in children, women of child-bearing age, and pregnant women, more detailed and systematic studies are required to define dosimetry, precision, minimal detection limits, and clinical utility in these populations. Dosimetry calculations on all tissues considered to be radiosensitive should be carried out according to NCRP and ICRP guidelines.
- Calibration of both instruments with standard reference materials or amputated limbs that parallel the mineral mass of the subjects to be measured is clearly needed. For the use of the K-line XRF instrument in postmenopausal osteoporotic women, the instrument should be calibrated with relevant standard reference materials or limbs (to reflect decreased mineral mass) obtained from postmenopausal women. For the L-line XRF instrument, studies are in progress to calibrate the instrument with the use of surgically amputated limbs from children.
- It is important to incorporate bone lead measurements in sensitive populations coupled to multiple outcome measures, and such outcome measures (biochemical, electrophysiologic, and neurobehavioral) can now be incorporated into cohort studies of infants, children, women of child-bearing age, pregnant women, and other adults with the L-line XRF method. For the K-line XRF instrument, similar or other outcome measures can be incorporated into cohort studies of occupationally exposed workers and postmenopausal women.

QUALITY ASSURANCE AND QUALITY CONTROL

Quality assurance includes all steps that are taken to ensure reliability of data, including use of scientifically and technically sound practices for the collection, transport, and storage of samples; laboratory analyses; and the recording, reporting, and interpretation of results (Friberg, 1988). Quality control focuses on the quality of laboratory results (Taylor, 1987) and consists of external quality control, which includes systems for objective monitoring of laboratory performance by an external laboratory, and internal quality control, which encompasses a defined set of procedures used by laboratory personnel to assess laboratory results as they are generated (Friberg, 1988). Those procedures lead to decisions about whether laboratory results are sufficiently reliable to be released. Auditing procedures are used to monitor sampling, transport of samples, and recording and reporting of data (Friberg, 1988); these procedures are intended to promote laboratory discipline and vigilance aimed at preventing errors, both inside and outside the laboratory (Taylor, 1987).

Statistical evaluation of laboratory data is essential for quality assurance and quality control; its primary objective is to assess analytic results for accuracy and precision. In this context, "precision" refers to the reproducibility of results, and "accuracy" specifies the validity of those results. Hence, precision is a measure of random errors of a method, and accuracy assesses systematic bias of the method. Random errors are always present and systematic errors sometimes occur. In the absence of systematic errors, the reproducibility of measurements places the ultimate limit on the confidence ranges that can be assigned to a set of analytic results. Similarly, the ability to detect systematic bias is limited by the analytic precision of the method: an inaccuracy of $\pm 20\%$ is unlikely to be detected when the reproducibility of the measurement is 20% or more.

Statistical evaluation of laboratory data is essential to detect systematic bias (Taylor, 1987). Whether a laboratory participates in a round-robin proficiency testing program or uses regression analyses, tests of homogeneity, or other statistical methods, the acceptability of the laboratory results is based on stringently defined methods for assessing potential error. A repeatedly observed deviation from an assumed value, whether statistically significant or not, should be a cause for concern and demand close scrutiny of laboratory results, even if no quality-control test rejected the laboratory (Taylor, 1987).

Inside the laboratory, the validity of each procedure must be established within a specific matrix, which replicates the matrix of biologic fluids or tissues to be analyzed. Internal laboratory checks include measurement of the stability of samples, extensive calibration and replication of samples measured in duplicate, determination of precision and accuracy testing with matrix-matched samples and standards, carrying through of blank solutions for analyses sequentially in every phase of analysis, and useful characterization of the range of linearity and variance of calibration curves over time (Vahter, 1982; Nieboer and Jusys, 1983; Stoeppler, 1983b). External assessments of laboratory performance are best carried out through a laboratory's participation in a well-organized, formalized interlaboratory proficiency testing program. Such a program should include the use of centrally prepared certified samples in which a specific characteristic has been measured with a reference method (Friberg, 1988). For example, a suitable proficiency testing program for measurement of lead in whole blood would involve the use of blood samples from cows fed an inorganic lead salt; the samples would be certified as to lead content with isotope-dilution mass spectroscopy at the National Institute of Standards and Technology.

A definitive method is one in which various characteristics are clearly defined and instrumental measurements can be performed with a high level of confidence (Cali and Reed, 1976). For lead in biologic fluids, the definitive method is isotope-dilution mass spectroscopy (IDMS). IDMS reaches a high level of accuracy, because manipulations are carried out on a weight basis. Measurement involves determinations of isotope ratios, not absolute determinations of individual isotopes; analytic precision of one part in 10^4-10^5 can be routinely obtained. Atomic-absorption spectroscopy and anodic stripping voltammetry both qualify as reference methods for measurement of lead in whole blood, because results obtained with them can be assessed precisely by or calibration against results of IDMS analyses (Stoeppler, 1983b). As acceptable concentrations of lead in blood and other biologic media become lower, the availability of standard reference materials and their wider distribution to laboratories become increasingly important.

Rapid advances in development of sophisticated instrumentation, increased awareness of background contamination of lead analyses outside and inside the laboratory, and the use of reference methods have contributed to laboratories capability to measure lower concentrations of lead and to measure lead with increased precision (Stoeppler, 1983b; Taylor, 1987; Friberg, 1988). Because lead is widely distributed in air, in dust, and in routinely used laboratory chemicals, the laboratory requirements for ultraclean or ultratrace analyses of lead, such as for lead in plasma, are extremely demanding; very few laboratories have ultraclean facilities and related instrumentation to permit accurate measurements of trace amounts of lead in biologic fluids (Patterson and Settle, 1976; Everson and Patterson, 1980). In an ideal world, ultraclean laboratory techniques would be applied in all laboratory procedures. In the practical world of clinical and epidemiologic studies, these exacting techniques are unrealistic. For example, excellence in laboratory standards is essential, but sampling conditions in the field, clinic, or hospital cannot be expected to be the same as ambient conditions in an ultraclean laboratory. Sampling under practical conditions is bound to involve some positive analytic error (Stoeppler, 1983b). However, the amount of lead introduced by ambient exposure of biologic fluids and by routinely used laboratory reagents must be known and assessed frequently to identify significant compromise within an analytic procedure (Rabinowitz and Needleman, 1982; Friberg, 1988). Temporal considerations are also important in the measurement of lead and biologic indicators of lead toxicity, because these substances might circulate in biologic fluids with specific and different patterns of ultradian (between an hour and a day) and circadian rhythmicity (Rabinowitz and Needleman, 1982).

Sample collection has the potential to account for the greatest amount of positive error during analyses of lead at low concentrations, in contrast with the smaller errors that are inherent in instrumentation (Nieboer and Jusys, 1983). In measurement of lead in whole blood, for example, blood for analysis is preferably obtained with venipuncture and less preferably with fingerstick. The choice of sampling technique is based on feasibility, setting (parent-child acceptance), and extent of training of the persons who are collecting the samples. Fingerstick or capillary sampling might be problematic because of skin contamination, contamination within capillary tubing, and inadequate heparinization of the sample (EPA, 1986a). Cleaning the skin thoroughly is more important than the choice between alcohol and a phosphate-containing soap as the cleanser. Overestimates of the concentration of lead in whole blood can occur through contamination of the skin or capillary tubes, and underestimates can occur through dilution of blood with tissue fluids caused by squeezing of the finger too rigorously. For those reasons, during NHANES II, only venous blood samples were obtained and included in the analysis of survey results (Annest et al., 1982).

It should be noted that the erythrocyte protoporphyrin (EP) test is an insensitive assay at blood lead concentrations below 50 μg/dL (Mahaffey and Annest, 1986); widespread use of properly obtained fingerstick samples appears to be necessary in screening large populations. The EP test has commonly been used in the past to screen large populations of children. From a public-health standpoint, it is desirable to obtain false-positive values with the new fingerstick techniques and then perform definitive venous sampling. Public-health officials should not rely on old EP methods, which have been shown to yield a false-negative rate of about 50% at blood lead concentrations less than 50 μg/dL (Mahaffey and Annest, 1986).

Plastic and glass laboratory equipment must be cleaned rigorously in an acid bath and then washed thoroughly in high-purity (18-megohm) water. Only laboratory reagents with known lead concentrations should be used, and these contributors to positive laboratory errors should be measured individually (Rabinowitz and Needleman, 1982). For ultra-trace analyses, doubly distilled reagents and highly purified materials are necessary (Everson and Patterson, 1980).

SUMMARY

Most clinical and epidemiologic research laboratories now involved in measuring lead in biologic materials use only a few well-studied analytic methods routinely. In coming years, most laboratories will probably retain these methods for measuring lead at the ever-lower guideline concentrations that are being promulgated, e.g., lead in whole blood at or below 10 μg/dL, the new CDC action level for childhood lead exposure. The routine methods used for such typical analyses as lead in whole blood are principally electrothermal or graphite-furnace atomic-absorption spectrometry (GF-AAS) and the electrochemical technique of

ASV. Other variations on AAS and electrochemistry are either passing out of use or increasing in use; they are not as popular with laboratories as those two. Both AAS and ASV are theoretically adequate for the new, more rigid performance and proficiency demands being placed on laboratories in light of lower body lead burdens and exposure and toxicity guidelines, provided that attention to rigid protocols is scrupulous.

It appears that blood lead measurements will continue to have an important place in the human toxicology of lead, primarily as an index of recent exposure. L-line XRF measurements of lead in tibial cortical bone of children, women of child-bearing age, pregnant women and other adults appear to have considerable dosimetric relevance for assessing lead stores that have accumulated over a lifetime. K-line XRF techniques appear to have substantial biologic relevance for assessing cumulative lead exposure in workers in lead-related industries and possibly for relating cumulative lead exposure to epidemiologic study of renal disease, hypertension, and osteoporosis. Wider application of both techniques, coupled to sensitive high-performance liquid chromatographic methods for assessing lead's inhibition of the heme biosynthetic pathway in worker populations, holds considerable promise for further delineating toxic effects of lead in sensitive populations at relatively low exposures.

The primary concern with current lead concentration measurements is over analytic error, rather than instrumental limitations—specifically caused by the introduction of lead into samples during collection, storage, treatment, and analysis. Contaminant lead, which is commonly not accurately quantified, increased measured lead concentrations. Consequently, reported lead concentrations in both biologic and environmental samples are often erroneously high, as demonstrated with intercalibrated studies that used trace-metal-clean techniques and rigorous quality-control and quality-assurance procedures.

Intercalibrations have demonstrated that many conventional instruments are capable of accurately and precisely measuring lead concentrations in biologic tissues and environmental samples. They include atomic-absorption spectrometers and anodic-stripping voltammeters, which are commonly used to measure lead concentrations in hospitals and commercial laboratories. Moreover, recent advances in instrumentation and methods have demonstrated that parts-per-trillion lead concentrations can be accurately and precisely measured with either of those instruments.

Noninvasive XRF methods are being used more for measuring lead in bone, where most of the body burden of lead accumulates. They include complementary methods: LXRF and KXRF. Developments in in vivo XRF analysis of human bone lead are occurring in parallel with increases in knowledge of two closely related subjects: the quantitative relation of lead exposure to bone lead concentrations and the quantitative relation of bone lead concentration, either total or compartment-specific, to either the extent of resorption into the circulation or the health risks associated with such resorption. The direction of future XRF methodologic developments and the ultimate potential of XRF methods for use in large-scale exposure screenings of diverse populations will be influenced by the answers to these questions:

- Are sensitivity limits of present XRF methods adequate for reliable measurement of bone lead concentrations that reflect unacceptable cumulative exposures and indicate some potential health risk associated with resorption into blood?
- Are thresholds of potential concern for adverse health effects of lead, when indexed as bone lead concentration, around or below the current measurement capability for in vivo lead quantitation?
- Are XRF systems likely to be useful only for screening high-exposure subjects, or can the potential for this in vivo determination extend to low-exposure subjects if instruments are refined?
- How reliably can the apparently distinct bone lead pools probed with LXRF vs. KXRF methods be related to past lead exposures that could lead to mobilization of lead from bone stores and increased future toxicity? Is tandem use of LXRF and KXRF measurements required in serial exposure screenings of sensitive populations for adequate risk assessment?
- Who are the most appropriate sensitive populations for XRF analysis? Young children? If so, at what age range? Women of childbearing age?
- To what purpose should XRF data be put? To determine that cumulative past exposure producing irreversible intoxication is a marker

of current adverse health effects? To determine the extent of future risk of endogenous toxicity associated with resorption of lead from bone into blood?

Technical problems with XRF methods appear to fall into three general categories, as follows:

- *Difficulty in separating instrument background from lead signal.* Research groups and instrument vendors use varied software to separate the lead signal from the instrument background. Some software is proprietary and has little publicly available documentation; sharing of applicable software could help to advance the field. Even some of the better methods for separating signal from background cannot identify a signal that is less than 2 times greater than background. Lead concentrations of 2-10 µg/g are the lowest that are quantifiable with XRF methods. Those concentrations are higher than bone lead concentrations of many adults and most children, but not higher than those of occupationally exposed people or people with atypically increased environmental lead exposures.
- *Variability between instruments.* No appropriate certified standards are available to provide a basis for obtaining consistent, accurate, and quantitative data. The presence of standards would make it possible to assess and improve the accuracy and precision of different instruments. The availability of standards would also improve the validity of quality-assurance and quality-control programs in use by various groups and permit comparative measurements between instruments.
- *Physiologic aspects of bone turnover.* Physiologic variability in lead concentration between and within various bones has not been explored with XRF methods. Bone turnover and remodeling are age- and sex-dependent, and that complicates interpretation among groups (especially young children) that have bone lead concentrations substantially below the limit of in vivo XRF methods now available.

Reported LXRF and KXRF systems appear to measure bone lead pools that differ in concentration, anatomic region, and toxicokinetic mobility. LXRF data might reflect a bone lead pool more labile than the purely cortical mineral lead fraction measured by the KXRF system. It would be premature to delineate the age-developmental time boundaries over which these two systems can be viewed as operating optimally. Consequently, it is appropriate to view the two approaches as providing complementary information. All such information might, in fact, be required to obtain a complete exposure profile and a comprehensive framework for assessment of health risks.

The definitive method of measuring lead concentration is still isotope-dilution thermal-ionization mass spectrometry (TIMS). Although TIMS is not appropriate for routine analyses, because of cost and complexity, it is becoming more useful for medical research as a result of continuing improvements in sample processing and sensitivity. The applicability of TIMS in medical research is also expanding with the development of stable-isotope tracer methods, which eliminate the need for radioisotopes in studies of lead metabolism.

Other types of inorganic mass spectrometers are becoming feasible. Major developments in inductively coupled plasma mass spectrometry have indicated that it will soon become common in hospitals and commercial laboratories, where it might be used for both lead concentration and isotopic-composition analyses of biologic tissues and environmental materials. Comparable advances in other types of mass spectrometry indicate that they could also soon be used to measure lead in solid materials; they include gas-discharge mass spectrometry, secondary-ion mass spectrometry, and laser-microprobe mass analysis.

Nuclear magnetic resonance spectroscopy is also potentially valuable for analysis of lead in biologic materials. Specifically, it might be used to measure intracellular and cellular lead, as well as calcium, zinc, iron, and other trace elements. However, applications of this technique will continue to be limited by the expense of the instrument and the technical expertise required to operate it.

Lead concentrations and isotopic compositions in biologic and environmental matrices can be accurately and precisely measured with existing instrumentation, notably atomic-absorption spectrometry and anodic-stripping voltammetry. The capabilities of those commonly used instruments now exceed most analytic requirements and are still being improved. The applicability of other instruments, which often provide complementary lead measurements, has been demonstrated within the last decade. It is anticipated that newer techniques (L-line and K-line XRF and inductively coupled plasma mass spectrometry) will soon be common in clinical and epidemiologic studies. Other types of mass spectrometry and nuclear magnetic resonance spectroscopy are also expected to be used in more specialized studies. The utility of all those types of analyses will continue to be limited by the degree of quality control and quality assurance used in sample collection, storage, and analysis.

As the focus of public-health officials has turned to lower exposures, the errors-in-variables problem has become more severe, and the need for more careful measurements has increased. For example, the standard error of the blind quality-control data in NHANES II was approximately 10% of the mean blood lead concentration. To achieve a similar signal-to-noise ratio in the data (and hence a similar reliability coefficient in epidemiologic correlations), NHANES III will need to reduce absolute measurement error by about a factor of 3. Similarly, as screening programs focus on lower exposures, the probability of misclassification increases, unless the measurement errors are reduced proportionally. Analyses of data from NHANES II, which had much better contamination control and laboratory technique than most screening programs, showed a significant risk of misclassification of a child's blood lead concentration as being above 30 μg/dL because of the analytic error (Annest et al., 1982). Reliable detection of blood lead concentrations of 10 μg/dL will require considerably more care and probably different methods from those now used.

6

Summary and Recommendations

Society and science are working hard to comprehend and respond to lead as a major, persisting public-health issue that is of particular relevance to what are termed sensitive populations (i.e., populations that are at special risk for the subtle adverse health effects of chronic low-dose lead exposure): infants, children, and pregnant women (as surrogates for fetuses). This chapter presents recommendations on such matters as sources and pathways of lead exposure, the environmental epidemiology of lead in sensitive populations, methods of assessing exposure to lead with reference to markers of both exposure and effect, and adverse health effects of lead.

SOURCES OF LEAD EXPOSURE

An understanding of lead exposure in sensitive populations requires knowledge of the possible sources of exposure. This is especially important for lead, because it is found widely throughout the environment. However, for sensitive populations, there are some sources that are more important than others. These include lead in paint, gasoline, drinking water, solder (used to solder joints in water-distribution systems and used in imported food packaging), and in imported pottery. As noted previously, there are some continued uses of lead arsenate in agriculture, and it may be a contaminant in some dietary supplements, such as calcium preparations. Finally, this report did not address occupational exposure, but this may also be an important indirect source of exposure for some children in families whose parents work in lead industries.

ADVERSE HEALTH EFFECTS OF LEAD

The committee has identified infants, children, and pregnant women (as surrogates for fetuses) as sensitive populations at risk for the subtle adverse health effects of chronic low-dose lead exposure. In addition, adults occupationally exposed to lead and others having potentially large exposures face the risk of various forms of lead toxicity, including risk of lead-induced increases in blood pressure. However, the most sensitive populations at special risk for lead toxicity are infants, children, and pregnant women (as surrogates for fetuses).

There has been a substantial change in our understanding of the health effects of lead since the mid-1970s. When the 1970s began, interest in lead was focused on symptomatic lead-poisoned children. Research beginning in the 1980's identified cognitive effects short of encephalopathy and numerous noncognitive end points of smaller and smaller lead exposures. Those changes have redefined the populations at risk and the risks themselves.

The 1980s also saw the advent of prospective studies examining the cognitive effects of prenatal and postnatal exposure to lead. Such studies have been undertaken in populations with much smaller exposures than were previously examined. Despite the lower power implied by the smaller exposure range and exposures closer to the measurement error in the analytic techniques used, studies by Bellinger, Dietrich, and Vimpani and their co-workers have all found decrements in the Bayley Scales of Infant Development (used to judge extent of growth and development) and other developmental tests for infants associated with prenatal, postnatal, or perinatal lead exposure. The blood lead concentrations in these studies were usually 5-15 μg/dL. Associations found in a study group consisting mostly of people with a history of alcohol and drug abuse were weaker or insignificant, possibly because of difficulties in detecting the effects of lead in a population exposed to other toxic agents. Even in that group, the trends generally suggested decreased performance with increased lead exposure.

Another development has been the use of end points other than full-scale IQ to measure the effects of lead on the central nervous system (CNS), such as teacher rating scales, reaction-time tests with an attention-span component, brain stem auditory evoked potentials, hearing thresholds, and other electroencephalographic measures. Some have found balance to be a sensitive early indicator of lead toxicity. The findings have added strength to the association with full-scale IQ, but it is noteworthy that the most consistent finding across all the studies of the CNS effects of lead is the association of increasing exposure with increasing reaction time, which apparently indicates an attention deficit. Similar effects have been noted in monkeys, again at lead concentrations of 5-15 μg/dL.

One complex of recently identified end points involves disturbances in the growth and development of the fetus and child. That identification has been accompanied by increased evidence of metabolic disturbances on the subcellular level that provide clues to the possible mechanisms of lead toxicity.

Recent studies indicate an inverse association of maternal lead exposure with birthweight or duration of pregnancy. For example, one study indicated that a change in maternal blood lead concentration from 4 to 8 μg/dL would be associated with a 0.6-week reduction in duration of pregnancy and a consequent 46-g reduction in birthweight. The joint teratogenic effects of alcohol, tobacco, and lead appear to be less than additive, and that could account for the lack of statistical significance in the finding of Ernhart and co-workers in a population with a history of alcohol abuse.

Effects of lead on postnatal growth and development have been reported. Shortness of heavily exposed children (mean blood lead, 57 μg/dL) was noted by Mooty in 1975. In 1986, Schwartz and co-workers reported an association between lead and stature, with a 3-cm decline in height at age 5 as blood lead concentrations increased from 5 to 25 μg/dL. Prospective studies have confirmed an association of lead exposure with retarded postnatal growth in infants and children. Lead has also been associated with a delay in the age at which a child first sits up. Again, the findings suggest significant differences between blood lead concentrations of 5 and 15 μg/dL.

Early epidemiologic assessments relied on traditional questionnaires (rote counting). If the majority of studies found an association, the effect was likely. The fact that the statistical analyses were highly sensitive to sample size was ignored, as were the sample sizes themselves. However, recent epidemiologic assessments have paid more attention to general trends and effect-size comparisons between studies and less attention to vote counting. Intuitively, five studies showing the same effect size are fairly convincing, whereas five studies with widely differing effect sizes are mostly suggestive of uncontrolled confounding. Meta-analysis is the common term for the more formal statistical methods for combining studies. Although the methods of meta-analysis must be applied cautiously, they promise an important improvement in the interpretation of the literature.

MARKERS OF LEAD EXPOSURE AND EFFECT

Lead has been shown to be a potent disturber of cellular calcium metabolism, preferentially binding to and activating or blocking calcium-binding proteins, such as calmodulin and calcitonin. Lead appears to produce increased intracellular calcium stores in every tissue studied, possibly with consequences for second-messenger functions. Recently, lead has been shown to activate protein kinase C at less than picomolar concentrations. No direct connection between those metabolic changes and disturbances in growth or cognitive functions has been established, but the existence of such changes adds considerable plausibility to the epidemiologic findings, particularly in view of the pervasive role of calcium in regulating cellular function.

Molecular biologic markers for assessing individual differences in responsiveness to lead exposure are of increasing interest, because of the well-known individual variations in susceptibility at a given blood lead concentration. Radioimmunoassays for marker proteins, such as the renal and brain lead-binding proteins that have been found in the blood and urine of rodents, are of great potential value in elucidating the underlying causative factors of susceptibility to lead toxicity at the molecular level.

The use of markers involves noninvasive or minimally invasive procedures. That is particularly important for gaining acceptance by public-health professionals and the public. It is clear that new marker techniques must be rapid, relatively simple, economically feasible, and associated with minimal health risks, if they are to gain widespread application.

The committee concludes that a number of questions concerning the systemic compartmental mobility of lead remain, especially those related to methods of monitoring for lead exposure.

- The committee recommends that further studies be done to determine the factors that influence movement of lead into bone and from bone to blood and other target tissues; the early-effect indicators in lead exposure monitoring, including research on lead-binding proteins and their use in monitoring for small exposures; and in vivo lead and calcium interactions.

TECHNIQUES TO MEASURE LEAD EXPOSURE AND EARLY TOXIC EFFECTS

Continuous measurement of exposure, allowing more detailed investigation of potential dose-response relations, has been introduced. Improved analytic techniques have reduced errors in the measurement of lead exposure, as well as of some covariates. Biologic markers of exposure—such as blood lead, urinary lead, and bone lead—have always involved measurement of total lead in epidemiologically and clinically relevant biologic media. The existence of specific biochemical forms of lead in accessible physiologic media, their role in lead toxicokinetics and toxicity, and their comparative diagnostic value in in situ biochemical behavior are important and perhaps require more attention in the future. In the near term, such questions would have to be addressed through in vivo metal research.

The potentially useful form-specific analysis in the near term for some segments of populations at risk has to do with speciation of inorganic versus biochemically stable organolead species in physiologic media with existing trace and ultratrace analytic methods. This approach is important for specific populations where environmental alkylation and accumulation occur or existence of alkyl lead forms would be problematic and where evidence exists of accumulation of different organolead species in tissues such as brain.

The current trend in measurements for lead dosimetry and effects is toward noninvasive or minimally invasive procedures for blood, urine, or bone. For bone, the anatomic location of the measurement is typically the finger, tibia, or calcaneus, because the bone in those areas is close to the skin and x-ray absorption in soft tissue is minimized. Those anatomic locations might also be physiologically important because of apparent differences between the residence time of lead in compact versus cancellous bone, which could influence calculations of internal lead doses from bone lead stores.

The regulatory and public-health implications of the advent of the techniques noted above concern the new ability to detect low-dose lead effects and to relate them more precisely to internal lead dosage. Further refinements and validation of the methods should permit societal decisions about low-dose lead in sensitive populations to be made on the basis of actual data, as opposed to calculated extrapolations, which can be based on uncorroborated assumptions. Clearly, the new methods have the potential to revolutionize public-health strategies in dealing with lead.

The committee concludes that, at the current blood lead concentrations of concern, accurate and precise blood lead values can be obtained with current techniques, given strict attention to contamination control and other principles of quality assurance and quality control (QA/QC). For the present and near future, blood lead values, rather than those of erythrocyte protoporphyrin, will be the primary screening tool to assess current lead exposure.

- The committee recommends that the optimal screening method should be venous sampling. However, the committee recognizes that initial screening of small children will involve capillary blood sampling with strict attention to principles of contamination control. Under the latter circumstances, a confirmatory followup measurement on children whose measurements exceed the latest Centers for Disease Control and Prevention guidelines should be carried out on a venous blood sample obtained by venipuncture.

The primary concerns associated with current measurements of lead concentrations in sensitive populations are unrecognized contamination and insufficient QA/QC. Several analytic techniques (atomic-absorption spectroscopy (AAS), anodic stripping voltammetry (ASV), and inductively coupled plasma mass spectrometry (ICP-MS)) are available for routine measurements of lead concentrations at parts-per-billion concentrations in clinical laboratories experienced in conducting those measurements regularly.

SUMMARY AND RECOMMENDATIONS 259

• The committee recommends the establishment of rigorous trace-metal cleanup techniques in sample collection, storage, and analysis in all clinical laboratories. The quality of those measurements should be documented with detailed QA/QC procedures, required participation in blind interlaboratory proficiency testing programs, and analysis of lead in blood with concurrent analyses of appropriate reference materials.

There is a need for stored samples and standard reference materials (bone, water, blood, urine, dust, soil, and paint) to assess laboratory precision and adherence to QA/QC principles. Aliquots of representative samples also need to be stored for future intercalibrations.

As the focus of epidemiologic studies has turned to smaller lead exposure, the problem of errors in the variables has become more severe, and the need for more careful measurements has markedly increased. Errors in measurement of variables other than lead similarly assume considerable importance. Before epidemiologic studies are begun, errors in variables (e.g., the standard deviation of the analytic measurement error in blood lead or IQ) must be systematically quantified.

For L-line and K-line x-ray fluorescence (XRF) instruments, standard reference materials are needed for intralaboratory and interlaboratory measurements and QA/QC assessments. Standard reference materials should be used to evaluate the counting time necessary to achieve a quantitative lead peak of, e.g., 10 ppm, with 95% confidence limits, under the same operating conditions used for patient measurements. If L-line or K-line instruments are proposed for use in women of childbearing age or pregnant women, the radiation risk to the human conceptus must be carefully quantified, according to NRC guidelines. Moreover, dosimetry measurements of both instruments should be calculated with strict adherence to National Council on Radiation Protection and Measurements procedures. The calculations should include determinations of absorbed- and scattered-dose rates, with radiation quality and tissue distribution weighting factors used for final calculations of the effective dose equivalent. The procedures should then provide confidence that risks associated with these techniques have been thoroughly examined.

• The committee recommends that federal agencies consider the need for further L-line and K-line XRF instrument development to decrease the counting time and enhance the detection limits without increasing radiologic risks. Both techniques may have future clinical utility to answer outstanding questions essential to protect the health of sensitive populations.

Federal agencies, which are mandated to clear new medical devices for clinical use, should be cognizant of the sequence of XRF instrument assessments that are necessary to ensure radiologic safety and clinical utility of these instruments in sensitive populations.

Recent advances in mass spectrometry have demonstrated the applicability of stable lead isotopes for investigating sources of environmental lead.

- The committee recommends that thermal-ionization mass spectrometry (TIMS) and ICP-MS be used to identify and trace unique sources of lead contamination that can be characterized by isotopic composition.

The same instrumentation could be used to investigate lead metabolism in humans with relatively small (microgram) amounts of a stable lead isotope tracer. The refinement of other analytic techniques that are still in development should be promoted for surface-area analyses (e.g., secondary-ion mass spectrometry, SIMS) and microanalyses (e.g., laser microprobe mass spectrometry, LAMMA).

In cells and tissues, lead has been shown to perturb the calcium messenger system. Although a direct connection between metabolic and dosimetric changes to disturbances in growth, development, vascular peripheral resistance, and cognitive function has yet to be fully established, the pervasive role of the calcium messenger in regulating diverse cellular functions provides considerable plausibility to epidemiologic findings. Given the inherent plausibility of those mechanistic and dosimetric observations, new initiatives and refinements in methods are needed to characterize further these and early toxic effects of lead on cells and tissues. Such new approaches and refinements in current techniques might become relevant in assessing lead exposure and toxic biochemical effects in sensitive populations.

The committee finds a need to measure the biologically active chemical species of lead that produce toxic effects at low doses and their relationship to lead binding in major intracellular compartments, such as lead-binding proteins, intranuclear inclusion bodies, mitochondria, and lysosomes. In addition, there is a need to understand further the mechanisms of low-level lead toxicity in target tissues, with particular emphasis on lead-induced changes in gene expression, calcium signaling, heme biosynthesis, and cellular energy production.

Current tissue-culture studies involve a degree of lead contamination in media and various reagents. As a result, even so-called untreated control cells can be perturbed, to an extent, by ambient lead in tissue-culture media.

• To understand further lead's mechanistic effects at the cellular level, the committee recommends that studies be conducted to explore the feasibility of applying ultraclean lead-free techniques to in vitro studies.

References

Ahlgren, L., K. Lidén, S. Mattsson, and S. Tejning. 1976. X-ray fluorescence analysis of lead in human skeleton in vivo. Scand. J. Work Environ. Health 2:82 86.
Ahlgren, L., B. Haeger-Aronsen, S. Mattsson, and A. Schütz. 1980. In vivo determination of lead in the skeleton after occupational exposure to lead. Br. J. Ind. Med. 37:109 113.
Alessio, L. 1988. Relationship between "chelatable level" and the indicators of exposure and effect in current and past occupational exposure. Sci. Total Environ. 71:293 299.
Alessio, L., P.A. Bertazzi, O. Monelli, and V. Foa. 1976a. Free erythrocyte protoporphyrin as an indicator of the biological effect of lead in adult males: II. Comparison between free erythrocyte protoporphyrin and other indicators of effects. Int. Arch. Occup. Environ. Health 37:89 105.
Alessio, L., P.A. Bertazzi, O. Monelli, and F. Toffoletto. 1976b. Free erythrocyte protoporphyrin as an indicator of the biological effect of lead in adult males: III. Behavior of free erythrocyte protoporphyrin in workers with past lead exposure. Int. Arch. Occup. Environ. Health 38:77 86.
Alessio, L., M.R. Castoldi, O. Monelli, F. Toffoletto, and C. Zocchetti. 1979. Indicators of internal dose in current and past exposure to lead. Int. Arch. Occup. Environ. Health 44:127 132.
Alexander, F.W., and H.T. Delves. 1981. Blood lead levels during pregnancy. Int. Arch. Occup. Environ. Health 48:35 39.
Alexander, F.W., H.T. Delves, and B.E. Clayton. 1973. The uptake

and excretion by children of lead and other contaminants. Pp. 319 331 in Proceedings of the International Symposium: Environmental Health Aspects of Lead, D. Barth, A. Berlin, R. Engel, P. Recht, and J. Smeets, eds. Luxembourg: Commission of the European Communities.

Alvares, A.P., A. Fischbein, S. Sassa, K.E. Anderson, and A. Kappas. 1976. Lead intoxication: Effects on cytochrome P-450-mediated hepatic oxidations. Clin. Pharmacol. Ther. 19:183 190.

Amitai, Y., J.W. Graef, M.J. Brown, R.S. Gerstle, N. Kahn, and R.E. Cochrane. 1987. Hazards of "deleading" homes of children with lead poisoning. Am. J. Dis. Child. 141:758 760.

Andersson-Engels, S., J. Johansson, S. Svanberg, and K. Svanberg. 1989. Fluorescence diagnosis and photochemical treatment of diseased tissue using lasers: Part I. Anal. Chem. 61:1367A 1373A.

Andersson-Engels, S., J. Johansson, S. Svanberg, and K. Svanberg. 1990. Fluorescence diagnosis and photochemical treatment of diseased tissue using lasers: Part II. Anal. Chem. 62:19A 27A.

Angerer, J., and K.H. Schaller, eds. 1988. Analyses of Hazardous Substances in Biological Materials, Vol. 2. Weinham, Germany: VCH Publishers. 252 pp.

Angle, C.R., and D.R. Kuntzelman. 1989. Increased erythrocyte protoporphyrins and blood lead—a plot study of childhood growth patterns. J. Toxicol. Environ. Health 26:149 156.

Angle, C.R., and M.S. McIntire. 1978. Low level lead and inhibition of erythrocyte pyrimidine nucleotidase. Environ. Res. 17:296 302.

Angle, C.R., and M.S. McIntire. 1979. Environmental lead and children: The Omaha study. J. Toxicol. Environ. Health 5:855 870.

Angle, C.R., and M.S. McIntire. 1982. Children, the barometer of environmental lead. Adv. Pediatr. 29:3 31.

Angle, C.R., M.S. McIntire, M.S. Swanson, and S.J. Stohs. 1982. Erythrocyte nucleotides in children—Increased blood lead and cytidine triphosphate. Pediatr. Res. 16:331 334.

Angle, C.R., A. Marcus, I.-H. Cheng, and M.S. McIntire. 1984. Omaha childhood blood lead and environmental lead: A linear total exposure model. Environ. Res. 35:160 170.

Annest, J.L. 1983. Trends in the blood lead levels of the US population: The Second National Health and Nutrition Examination Survey

(NHANES II) 1976 1980. Pp. 33 58 in Lead Versus Health, M. Rutter and R.R. Jones, eds. New York: John Wiley & Sons.

Annest, J.L., and K. Mahaffey. 1984. Blood lead levels for persons ages 6 months 74 years, United States: 1976 80. Vital and Health Statistics, Series 11, No. 233. DHHS Publ. No. (PHS) 84-1683. Washington, D.C.: National Center for Health Statistics.

Annest, J.L., K.R. Mahaffey, D.H. Cox, and J. Roberts. 1982. Blood lead levels for persons 6 months 74 years of age: United States, 1976-80. Pp. 1 23 in Advance Data from Vital and Health Statistics, No. 79. DHHS Publ. No. (PHS) 82-1250. Washington, D.C.: National Center for Health Statistics.

Annest, J.L., J.L. Pirkle, D. Makuc, J.W. Neese, D.D. Bayse, and M.G. Kovar. 1983. Chronological trend in blood lead levels between 1976 and 1980. N. Engl. J. Med. 308:1373 1377.

Apostoli, P., G. Maronelli, and R. Micciolo. 1992. Is hypertension a confounding factor in the assessment of blood lead reference values. Sci. Total Environ. 120:127 134.

Araki, S., and K. Ushio. 1982. Assessment of the body burden of chelatable lead: A model and its application to lead workers. Br. J. Ind. Med. 39:157 160.

Armstrong, R., D.R. Chettle, M.C. Scott, L.J. Somervaille, and M. Pendlington. 1992. Repeated measurements of tibia lead concentrations by in vivo x ray fluorescence in occupational exposure. Br. J. Ind. Med. 49:14 16.

Arvik, J.H., and R.L. Zimdahl. 1974. Barriers to the foliar uptake of lead. J. Environ. Qual. 3:369 373.

Aslam, M., S.S. Davis, and M.A. Healy. 1979. Heavy metals in some Asian medicines and cosmetics. Public Health 93:274 284.

Assennato, G., C. Paci, R. Molinini, R.G. Candela, M.E. Baser, B.M. Altamura, and R. Giorgino. 1986. Sperm count suppression without endocrine dysfunction in lead-exposed men. Arch. Environ. Health 41:387 390.

Astrin, K.H., D.F. Bishop, J.G. Wetmur, B. Kaul, B. Davidow, and R.J. Desnick. 1987. δ-Aminolevulinic acid dehydratase isozymes and lead toxicity. Ann. N.Y. Acad. Sci. 514:23 29.

ASTPHLD (Association of State and Territorial Public Health Laboratory Directors). 1991. Proceedings of the First National Conference on Laboratory Issues in Childhood Lead Poisoning Prevention, Oct. 31 Nov. 2, 1991, Columbia, Md. Washington, D.C.: Association of State and Territorial Public Health Laboratory Directors.

Atchison, W.D., and T. Narahashi. 1984. Mechanism of action of lead on neuromuscular junctions. NeuroToxicology 5:267 282.

ATSDR (Agency for Toxic Substances and Disease Registry). 1988. The Nature and Extent of Lead Poisoning in Children in the United States: A Report to Congress. Atlanta, Ga.: U.S. Department of Health and Human Services.

Aub, J.C., L.T. Fairhall, A.S. Minot, P.P. Reznikoff, and A. Hamilton. 1926. Lead Poisoning. Medical Monographs, Vol. 7. Baltimore, Md.: Williams & Wilkins.

Baer, R.D., J. Garcia de Alba, L.M. Cueto, A. Ackerman, and S. Davison. 1987. Lead as a Mexican folk remedy: Implications for the United States. Pp. 111 119 in Childhood Lead Poisoning: Current Perspectives. Proceedings of a National Conference, December 1 3, 1987. Washington, D.C.: Bureau of Maternal and Child Health, Health Resources and Services Administration, U.S. Department of Health and Human Services.

Baghurst, P.A., A.J. McMichael, G.V. Vimpani, E.F. Robertson, P.D. Clark, and N.R. Wigg. 1987a. Determinants of blood lead concentrations of pregnant women living in Port Pirie and surrounding areas. Med. J. Aust. 146:69 73.

Baghurst, P.A., E.F. Robertson, A.J. McMichael, G.V. Vimpani, N.R. Wigg, and R.R. Roberts. 1987b. The Port Pirie cohort study: Lead effects on pregnancy outcome and early childhood development. NeuroToxicology 8:395 401.

Baghurst, P.A., A.J. McMichael, N.R. Wigg, G.V. Vimpani, E.F. Robertson, R.R. Roberts, and S.L. Tong. 1992. Environmental exposure to lead and children's intelligence at the age of seven years: The Port Pirie Cohort Study. N. Engl. J. Med. 327:1279 1284.

Baker, E.L., Jr., D.S. Folland, T.A. Taylor, M. Frank, W. Peterson, G. Lovejoy, D. Cox, J. Housworth, and P.J. Landrigan. 1977. Lead poisoning in children of lead workers: Home contamination with industrial dust. N. Engl. J. Med. 296:260 261.

Baker, E.L., Jr., P.J. Landrigan, A.G. Barbour, D.H. Cox, D.S. Folland, R.N. Ligo and J. Throckmorton. 1979. Occupational lead poisoning in the United States: Clinical and biochemical findings related to blood lead levels. Br. J. Ind. Med. 36:314 322.

Barltrop, D. 1969. Transfer of lead to the human foetus. Pp. 135 151 in Mineral Metabolism in Pediatrics, D. Barltrop and W.L. Burland, eds. Philadelphia: Davis.

Barltrop, D. 1972. Children and enviromental lead. Pp. 52 60 in Lead in the Environment: Proceedings of a Conference, P. Hepple, ed. London: Institute of Petroleum.

Barltrop, D., and F. Meek. 1979. Effect of particle size on lead absorption from the gut. Arch. Environ. Health 34:280 285.

Barnes, I.L., J.W. Gramlich, M.G. Diaz, and R.H. Brill. 1978. The possible change of lead isotope ratios in the manufacture of pigments: A fractionation experiment. Adv. Chem. Ser. 171 (Archaeol. Chem. 2):273 277.

Barrie, L.A., and R.J. Vet. 1984. The concentration and deposition of acidity, major ions and trace metals in the snowpack of the eastern Canadian Shield during the winter of 1980 1981. Atmos. Environ. 18:1459 1470.

Barry, P.S.I. 1975. A comparison of concentrations of lead in human tissues. Br. J. Ind. Med. 32:119 139.

Barry, P.S.I. 1981. Concentrations of lead in the tissues of children. Br. J. Ind. Med. 38:61 71.

Barsony, J. and S.J. Marx. 1988. Ongoing protein synthesis needed for 1,25-$(OH)_2$-D_3 mediated increase of cyclic GMP in human skin fibroblasts. FEBS Lett. 235:207 210.

Barton, J.C. 1989. Retention of radiolead by human erythrocytes in vitro. Toxicol. Appl. Pharmacol. 99:314 322.

Battistuzzi, G., R. Petrucci, L. Silvagni, F.R. Urbani, and S. Caiola. 1981. Delta-aminolevulnate dehydrase: A new genetic polymorphism in man. Ann. Hum. Genet. 45:223 229.

Batuman, V., E. Landy, J.K. Maesaka, and R.P. Wedeen. 1983. Contribution of lead to hypertension with renal impairment. N. Engl. J. Med. 309:17 21.

Beauchemin, D., M.E. Bednas, S.S. Berman, J.W. McLaren, K.W.M. Siu, and R.E. Sturgeon. 1988a. Identification and quantitation of arsenic species in a dogfish muscle reference material for trace elements. Anal. Chem. 60:2209 2212.

Beauchemin, D., J.W. McLaren, S.N. Willie, and S.S. Berman. 1988b. Determination of trace metals in marine biological reference materials by inductively coupled plasma mass spectrometry. Anal. Chem. 60:687 691.

Beauchemin, D., J.W. McLaren, and S.S. Berman. 1988c. Use of external calibration for the determination of trace metals in biological materials by inductively coupled plasma mass spectrometry. J. Anal. At. Spectrom. 3:775 780.

Bellinger, D., and H.L. Needleman. 1992. Neurodevelopmental effects of low-level lead exposure in children. Pp. 191 208 in Human Lead Exposure, H.L. Needleman, ed. Boca Raton, Fla.: CRC Press.

Bellinger, D.C., H.L. Needleman, A. Leviton, C. Waternaux, M.B. Rabinowitz, and M.L. Nichols. 1984a. Early sensory-motor development and prenatal exposure to lead. Neurobehav. Toxicol. Teratol. 6:387 402.

Bellinger, D., H.L. Needleman, R. Bromfield, and M. Mintz. 1984b. A followup study of the academic attainment and classroom behavior of children with elevated dentine lead levels. Biol. Trace Elem. Res. 6:207 223.

Bellinger, D., A. Leviton, H.L. Needleman, C. Waternaux, and M. Rabinowitz. 1986a. Low-level lead exposure and infant development in the first year. Neurobehav. Toxicol. Teratol. 8:151 161.

Bellinger, D., A. Leviton, M. Rabinowitz, H. Needleman, and C. Waternaux. 1986b. Correlates of low-level lead exposure in urban children at 2 years of age. Pediatrics 77:826 833.

Bellinger, D., A. Leviton, C. Waternaux, H. Needleman, and M. Rabinowitz. 1987. Longitudinal analyses of prenatal and postnatal lead exposure and early cognitive development. N. Engl. J. Med. 316:1037 1043.

Bellinger, D., A. Leviton, C. Waternaux, H. Needleman, and M. Rabinowitz. 1988. Low-level lead exposure, social class, and infant development. Neurotoxicol. Teratol. 10:497 503.

Bellinger, D., J. Sloman, A. Leviton, C. Waternaux, H. Needleman, and M. Rabinowitz. 1989a. Environmental exposure to lead and cognitive deficits in children [letter]. N. Engl. J. Med. 320:595 596.

Bellinger, D., A. Leviton, C. Waternaux, H. Needleman, and M. Rabinowitz. 1989b. Low-level lead exposure and early development in socioeconomically advantaged urban infants. Pp. 345 356 in Lead Exposure and Child Development: An International Assessment, M.A. Smith, L.D. Grant, and A.I. Sors, eds. Dordrecht, The Netherlands: Kluwer Academic.

Bellinger, D., A. Leviton, and J. Sloman. 1990. Antecedents and correlates of improved cognitive performance in children exposed in utero to low levels of lead. Environ. Health Perspect. 89:5 11.

Bellinger, D., J. Sloman, A. Leviton, M. Rabinowitz, H.L. Needleman, and C. Waternaux. 1991a. Low-level lead exposure and children's cognitive function in the preschool years. Pediatrics 87:219 227.

Bellinger, D., A Leviton, M. Rabinowitz, E. Allred, H. Needleman, and S. Schoenbaum. 1991b. Weight gain and maturity in fetuses exposed to low levels of lead. Environ. Res. 54:151 158.

Bellinger, D.C., K.M. Stiles, and H.L. Needleman. 1992. Low-level lead exposure, intelligence, and academic achievement: A long-term follow-up study. Pediatrics 90:855 861.

Bellinger, D., H. Hu, L. Titlebaum, and H. Needleman. In press. Attentional correlates of dentin and bone lead levels in adolescents. Arch. Environ. Health.

Benignus, V.A., D.A. Otto, K.E. Muller, and K.J. Seiple. 1981. Effects of age and body burden of lead on CNS function in young children: II. EEG spectra. Electroencephalogr. Clin. Neurophysiol. 52:240 248.

Bergomi, M., P. Borella, G. Fantuzzi, G. Vivoli, N. Sturloni, G. Cavazzuti, A. Tampieri, and P. Tartoni. 1989. Relationship between lead exposure indicators and neuropsychological performance in children. Dev. Med. Child Neurol. 31:181 190.

Berman, S.S., K.W.M. Siu, P.S. Maxwell, D. Beauchemin, and V.P. Clancy. 1989. Marine biological reference materials for methylmercury analytical methodologies used in certification. Fresenius Z. Anal. Chem. 333:641 644.

Bernard, B.P., and C.E. Becker. 1988. Environmental lead exposure and the kidney. Clin. Toxicol. 26:1 34.

Biggins, P.D.E., and R.M. Harrison. 1979. Atmospheric chemistry of automotive lead. Environ. Sci. Technol. 13:558 565.

Binder, S., D. Sokal, and D. Maughan. 1986. Estimating soil ingestion: The use of tracer elements in estimating the amount of soil ingested by young children. Arch. Environ. Health 41:341 345.

Blake, K.C.H. 1976. Absorption of ^{203}Pb from gastrointestinal tract of man. Environ. Res. 11:1 4.

Blake, K.C.H., and M. Mann. 1983. Effect of calcium and phosphorus on the gastrointestinal absorption of ^{203}Pb in man. Environ. Res. 30:188 194.

Blake, K.C.H., G.O. Barbezat, and M. Mann. 1983. Effect of dietary constituents on the gastrointestinal absorption of ^{203}Pb in man. Environ. Res. 30:182 187.

Blanksma, L.A., H.K. Sachs, E.F. Murray, and M.J. O'Connell. 1970. Failure of the urinary δ-aminolevulinic acid test to detect pediatric lead poisoning. Am. J. Clin. Pathol. 53:956 962.

Bloch, P., G. Garavaglia, G. Mitchell, and I.M. Shapiro. 1977. Measurement of lead content of children's teeth in situ by x-ray fluorescence. Phys. Med. Biol. 22:56 63.

Bonucci, E., and G. Silvestrini. 1988. Ultrastructural studies in experimental lead intoxication. Contrib. Nephrol. 64:93 101.

Bogden, J.D., S.B. Gertener, F.W. Kemp, R. McLeod, K.S. Bruening, and H.R. Chung. 1991. Dietary lead and calcium: Effects on blood pressure and renal neoplasia in Wistar rats. J. Nutr. 121:718 728.

Borella, P., P. Picco, and G. Masellis. 1986. Lead content in abortion material from urban women in early pregnancy. Int. Arch. Occup. Environ. Health 57:93 99.

Bornschein, R.L., P. Succop, K.N. Dietrich, C.S. Clark, S. Que Hee, and P.B. Hammond. 1985. The influence of social and environmental factors on dust lead, hand lead, and blood lead levels in young children. Environ. Res. 38:108 118.

Bornschein, R.L., P.A. Succop, K.M. Krafft, C.S. Clark, B. Peace, and P.B. Hammond. 1987. Exterior surface dust lead, interior house dust lead and childhood lead exposure in an urban environment. Trace Substances Environ. Health 20:322 332.

Bornschein, R.L., J. Grote, T. Mitchell, P.A. Succop, K.N. Dietrich, K.M. Krafft, and P.B. Hammond. 1989. Effects of prenatal lead exposure on infant size at birth. Pp. 307 319 in Lead Exposure and Child Development: An International Assessment, M.A. Smith, L.D. Grant, and A.I. Sors, eds. Dordrecht, The Netherlands: Kluwer Academic.

Boscolo, P., and M. Carmignani. 1988. Neurohumoral blood pressure regulation in lead exposure. Environ. Health Perspect. 78:101 106.
Bose, A., K. Vashistha, and B.J. O'Loughlin. 1983. Azarcon por empacho—Another cause of lead toxicity. Pediatrics 72:106 108.
Boutron, C.F., and C.C. Patterson. 1983. The occurrence of lead in Antarctic recent snow, firn deposited over the last two centuries and prehistoric ice. Geochim. Cosmochim. Acta 47:1355 1368.
Boutron, C.F., and C.C. Patterson. 1986. Lead concentration changes in Antarctic ice during the Wisconsin/Holocene transition. Nature 323:222 225.
Boyle, E.A., S.D. Chapnick, G.T. Shen, and M.P. Bacon. 1986. Temporal variability of lead in the western North Atlantic. J. Geophys. Res. C: Oceans 91:8573 8593.
Breart, G., Y. Rabarison, P.F. Plouin, C. Sureau, and C. Rumeau-Rouquette. 1982. Risk of fetal growth retardation as a result of maternal hypertension. Preparation to a trial of antihypertensive drugs. Dev. Pharmacol. Ther. 4(Suppl.):116 123.
Bressler, J.P., and G.W. Goldstein. 1991. Mechanisms of lead neurotoxicity. Biochem. Pharmacol. 41:479 874.
Brockhaus, A., W. Collet, R. Dolgner, R. Engelke, U. Ewers, I. Freier, E. Jermann, S. Krämer, U. Krämer, N. Manojlovic, M. Turfeld, and G. Winneke. 1988. Exposure to lead and cadmium of children living in different areas of North-West Germany: Results of biological monitoring studies 1982-1986. Int. Arch. Occup. Environ. Health 60:211 222.
Brown, R.M., and C.J. Pickford. 1985. Inductively coupled plasma-mass spectrometry: An individual assessment of the VG isotopes Plasmaquad. INIS Atomindex 16(Abstract 070138, Report AERE-M-3462).
Bruland, K.W. 1983. Trace elements in sea-water. Pp. 157 220 in Chemical Oceanography, Vol. 8, J.P. Riley and R. Chester, eds. London: Academic.
Brunekreef, B.D. 1984. The relationship between air lead and blood lead in children: A critical review. Sci. Total Environ. 38:79 123.
Buc, H.A., and J.C. Kaplan. 1978. Red-cell pyrimidine 5'-nucleotidase and lead poisoning. Clin. Chim. Acta 87:49 55.

Buchet, J.P., H. Roels, G. Hubermont, and R. Lauwerys. 1978. Placental transfer of lead, mercury, cadmium, and carbon monoxide in women: II. Influence of some epidemiological factors on the frequency distributions of the biological indices in maternal and umbilical cord blood. Environ. Res. 15:494 503.

Buchet, J.P., H. Roels, A. Bernard, Jr., and R. Lauwerys. 1980. Assessment of renal function of workers exposed to inorganic lead, cadmium or mercury vapor. J. Occup. Med. 22:741 750.

Bushnell, P.J., and R.E. Bowman. 1979. Persistence of impaired reversal learning in young monkeys exposed to low levels of dietary lead. J. Toxicol. Environ. Health 5:1015 1023.

Byers, R.K., and E.E. Lord. 1943. Late effects of lead poisoning on mental development. Am. J. Dis. Child. 66:471 494.

Calabrese, E.J., R. Barnes, E.J. Stanek, III, H. Pastides, C.E. Gilbert, P. Veneman, X.R. Wang, A. Lasztity, and P.T. Kostecki. 1989. How much soil do young children ingest: An epidemiologic study. Regul. Toxicol. Pharmacol. 10:123 137.

Calderón, V., C. Hernández-Luna, D. Sáenz, and M. y Maldonado. 1992. Biochemical Mechanisms Against the Toxicological Effects of Lead: A Study of Children in Mexico City [in Spanish]. Paper presented at the Fourth Annual Meeting of the International Society for Environmental Epidemiology and the International Society of Exposure Analysis, Aug. 26 29, 1992, Cuernavaca, Morelos, Mexico.

Cali, J.P., and W.P. Reed. 1976. The role of the National Bureau of Standards: Standard reference materials in accurate trace analysis. Pp. 41 63 in Accuracy in Trace Analysis: Sampling, Sample Handling, and Analysis, Vol. 1, P.D. LaFleur, ed. National Bureau of Standards Special Publ. 422. Washington, D.C.: U.S. Department of Commerce.

Camerlynck, R., and L. Kiekens. 1982. Speciation of heavy metals in soils based on charge separation. Plant Soil 68:331 339.

Campbell, M.J., and H.T. Delves. 1989. Accurate and precise determination of lead isotope ratios in clinical and environmental samples using inductively coupled plasma source mass spectrometry. J. Anal. At. Spectrom. 4:235 236.

Campbell, B.C., P.A. Meredith, M.R. Moore, and W.S. Watson. 1984. Kinetics of lead following intravenous administration in man. Toxicol. Lett. 21:231 235.

Cantarow, A., and M. Trumper. 1944. Lead Poisoning. Baltimore: Williams & Wilkins. 264 pp.
Cantor, K.P., J.M. Sontag, and M.F. Held. 1986. Patterns of mortality among plumbers and pipefitters. Am. J. Ind. Med. 10:73 89.
Capar, S.G. 1991. Analytical method aspects of assessing dietary intake of trace elements. Pp. 181 195 in Biological Trace Element Research. ACS Symposium Series 445, K.S. Subramanian, G.V. Iyengar, and K. Okamoto, eds. Washington, D.C.: American Chemical Society.
Caplun, E., D. Pettit, and E. Picciotto. 1984. Lead in petrol. Endeavor 8:135 144.
Carmignani, M., P. Boscolo, G. Ripanti, and V.N. Finelli. 1983. Effects of chronic exposure to cadmium and/or lead on some neurohumoral mechanisms regulating cardiovascular function in the rat. Pp. 557 560 in International Conference on Heavy Metals in the Environment, Vol. 1. Edinburgh: CEP Consultants.
Carton, J.A., J.A. Maradona, and J.M. Arribas. 1987. Acute-subacute lead poisoning. Clinical findings and comparative study of diagnostic tests. Arch. Intern. Med. 147:697 703.
Cavalleri, A., C. Minoia, L. Pozzoli, and A. Baruffini. 1978. Determination of plasma lead levels in normal subjects and in lead-exposed workers. Br. J. Ind. Med. 35:21 26.
CDC (Centers for Disease Control). 1975. Increased Lead Absorption and Lead Poisoning in Young Children. Atlanta, Ga.: CDC, U.S. Department of Health, Education, and Welfare.
CDC (Centers for Disease Control). 1978. Preventing Lead Poisoning in Young Children: A Statement by the Centers for Disease Control. DHEW Publ. No. 00-2629. Atlanta, Ga.: CDC, U.S. Department of Health and Human Services.
CDC (Centers for Disease Control). 1982. Annual Summary 1981: Reported morbidity and mortality in the United States. MMWR 30(54).
CDC (Centers for Disease Control). 1983a. Folk remedy-associated lead poisoning in Hmong children—Minnesota. MMWR 32:555 556.
CDC (Centers for Disease Control). 1983b. Lead poisoning from Mexican folk remedies. MMWR 32:554 555.
CDC (Centers for Disease Control). 1985. Preventing Lead Poisoning in Young Children: A Statement by the Centers for Disease Control,

January 1985, DHHS Publ. No. 99-2230. Atlanta, Ga: CDC, U.S. Department of Health and Human Services.

CDC (Centers for Disease Control). 1986a. East Helena, Montana, child lead study, summer 1983: Final report, July 1986, Atlanta, Ga.: CDC, U.S. Department of Health and Human Services.

CDC (Centers for Disease Control). 1986b. Final Report: Kellogg Revisited—1983. Childhood Blood Lead and Environmental Status Report. Joint Report with Idaho Department of Health and Welfare and U.S. Environmental Protection Agency. Atlanta, Ga.: CDC, U.S. Department of Health and Human Services.

CDC (Centers for Disease Control). 1991. Preventing Lead Poisoning in Young Children: A Statement by the Centers for Disease Control—October 1991. Atlanta, Ga.: CDC, U.S. Department of Health and Human Services.

Chai, S., and R.C. Webb. 1988. Effects of lead on vascular reactivity. Environ. Health Perspect. 78:85 89.

Chamberlain, A.C. 1983. Effect of airborne lead on blood lead. Atmos. Environ. 17:677 692.

Chamberlain, A.C. 1985. Prediction of response of blood lead to airborne and dietary lead from volunteer experiments with lead isotopes. Proc. R. Soc. London Ser. B 224:149 182.

Chamberlain, A.C., M.J. Heard, P. Little, D. Newton, A.C. Wells, and R.D. Wiffen. 1978. Investigations into lead from motor vehicles, Rep. AERE-R9198. Harwell, U.K.: United Kingdom Atomic Energy Authority.

Chamberlain, A.C., M.J. Heard, P. Little, and R.D. Wiffen. 1979. The dispersion of lead from motor exhausts. Philos. Trans. R. Soc. London. Ser. A 290:577 589.

Charney, E., B. Kessler, M. Farfel, and D. Jackson. 1983. Childhood lead poisoning: A controlled trial of the effect of dust-control measures on blood lead levels. N. Engl. J. Med. 309:1089 1093.

Chartsias, B., A. Colombo, D. Hatzichristidis, and W. Leyendecker. 1986. The impact of gasoline lead on man's blood lead: First results of the Athens lead experiment. Sci. Total Environ. 55:275 282.

Chassard-Bouchaud, C. 1987. Ion microscopes and microprobes in marine pollution research. Anal. Chim. Acta 195:307 315.

Chaube, S., C.A. Swinyard, and H. Nishimura. 1972. A quantitative

study of human embryonic and fetal lead with considerations of maternal-fetal lead gradients and the effect of lead on human reproduction. Teratology 5:253.
Chettle, D.R., M.C. Scott, and L.J. Somervaille. 1989. Improvements in the precision of in vivo bone lead measurements. Phys. Med. Biol. 34:1295 1300.
Chettle, D.R., M.C. Scott, and L.J. Somervaille. 1991. Lead in bone: Sampling and quantitation using K x-rays excited by ^{109}Cd. Environ. Health Perspect. 91:49 55.
Chia, K.S., C.N. Ong, H.Y. Ong, and G. Endo. 1989. Renal tubular function of workers exposed to low levels of cadmium. Br. J. Ind. Med. 46:165 170.
Chisolm, J.J., Jr. 1962. Aminoaciduria as a manifestation of renal tubular injury in lead intoxication and a comparison with patterns of aminoaciduria seen in other diseases. J. Pediatr. 60:1 17.
Chisolm, J.J., Jr. 1965. Chronic lead intoxication in children. Dev. Med. Child Neurol. 7:529 536.
Chisolm, J.J., Jr. 1968. The use of chelating agents in the treatment of acute and chronic lead intoxication in childhood. J. Pediatr. 73:1 38.
Chisolm, J.J., Jr. 1981. Dose-effect relationship for lead in young children: Evidence in children for interactions among lead, zinc and iron. Pp. 1 7 in Environmental Lead. Proceedings of the Second International Symposium on Environmental Lead Research, D.R. Lynam, L.G. Piantanida, and J.F. Cole, eds. New York: Academic.
Chisolm, J.J., Jr., and D. Barltrop. 1979. Recognition and management of children with increased lead absorption. Arch. Dis. Child. 54:249 262.
Chisolm, J.J., Jr., and H.E. Harrison. 1956. The exposure of children to lead. Pediatrics 18:943 958.
Chisolm, J.J., Jr., M.B. Barrett, and H.V. Harrison. 1975. Indicators of internal dose of lead in relation to derangement in heme synthesis. Johns Hopkins Med. J. 137:6 12.
Chisolm, J. J., Jr., E.D. Mellits, and M. B. Barrett. 1976. Interrelationships among blood lead concentration, quantitative daily ALA-U and urinary lead output following calcium EDTA. Pp. 416 433 in Proceedings of Third Meeting of the Subcommittee on the Toxicolo-

gy of Metals under the Permanent Commission and International Association on Occupational Health, G.F. Nordberg, ed. Amsterdam: Elsevier.

Chisolm, J.J., Jr., E.D. Mellits, and S.A. Quaskey. 1985. The relationship between the level of lead absorption in children and the age, type, and condition of housing. Environ. Res. 38:31 45.

Chisolm, J.C., S. Kim, and A.H. Tashjian, Jr. 1988. Modulation by 1,25-dihydroxycholecalciferol of the acute change in cytosolic free calcium induced by thyrotropin-releasing hormone in GH4C1 pituitary cells. J. Clin. Invest. 81:661 668.

Chmiel, K.M., and R.M. Harrison. 1981. Lead content of small mammals at a roadside site in relation to the pathways of exposure. Sci. Total. Environ. 17:145 154.

Chow, T.J., and M.S. Johnstone. 1965. Pb isotopes in gasoline and aerosols of the Los Angeles basin. Science 147:502 503.

Chow, T.J., C.B. Snyder, and J.L. Earl. 1975. Isotope ratios of lead as pollutant source indicators. Pp. 95 108 in Isotope Ratios of Lead as Pollutant Source and Behavior Indicators. Proceedings of a symposium, Vienna, Nov. 18 22, 1974, B. Kaufmann, ed. Vienna: International Atomic Energy Agency.

Christoffersson, J.O., A. Schütz, L. Ahlgren, B. Haeger-Aronsen, S. Mattsson, and S. Skerfving. 1984. Lead in finger-bone analyzed in vivo in active and retired lead workers. Am. J. Ind. Med. 6:447 457.

Christoffersson, J.O., L. Ahlgren, A. Schütz, S. Skerfving, and S. Mattsson. 1986. Decrease of skeletal lead levels in man after end of occupational exposure. Arch. Environ. Health 41:312 318.

Clark, C.S., R.L. Bornschein, P. Succop, S.S. Que Hee, P.B. Hammond, and B. Peace. 1985. Condition and type of housing as an indicator of potential environmental lead exposure and pediatric blood lead levels. Environ. Res. 38:46 53.

Clark, C.S., R.L. Bornschein, P. Succop, P.B. Hammond, B. Peace, K. Krafft, and K. Dietrich. 1987. Pathways to elevated blood lead and their importance in control strategy development. Pp. 159 161 in International Conference on Heavy Metals in the Environment, Vol. 1, S.E. Lindberg and T.C. Hutchinson, eds. Edinburgh: CEP Consultants.

Clarkson, T.W., L. Friberg, G.F. Nordberg, and P.R. Sager, eds.

1988. Biological Monitoring of Toxic Metals. Rochester Series on Environmental Toxicity. New York: Plenum Press. 686 pp.

Clausing, P., B. Brunekreef, and J.H. van Wijnen. 1987. A method for estimating soil ingestion by children. Int. Arch. Occup. Environ. Health 59:73 82.

Cohen, N. 1970. The retention and distribution of lead-210 in the adult baboon [dissertation]. New York: New York University School of Engineering and Science.

Cook, J.D. 1990. Adaptation in iron metabolism. Am. J. Clin. Nutr. 51:301 308.

Cook, L.R., C.R. Angle, and S.J. Stohs. 1986. Erythrocyte arginase, pyrimidine 5'-nucleotidase (P5N) and deoxypyrimidine 5'-nucleotidase (dP5N)' as indices of lead exposure. Br. J. Ind. Med. 43:387 390.

Cook, L.R., C.R. Angle, C.E. Kubitschek, and S.J. Stohs. 1987. Prediction of blood lead by HPLC assay of erythrocyte pyrimidine-5'-nucleotidase. J. Anal. Toxicol. 11:39 42.

Cookman, G.R., W.B. King, and C.M. Regan. 1987. Chronic low-level lead exposure impairs embryonic to adult conversion of the neural cell adhesion molecule. J. Neurochem. 49:399 403.

Cooney, G.H., A. Bell, W. McBride, and C. Carter. 1989a. Neurobehavioural consequences of prenatal low level exposures to lead. Neurotoxicol. Teratol. 11:95 104.

Cooney, G.H., A. Bell, W. McBride, and C. Carter. 1989b. Low-level exposures to lead: The Sydney lead study. Dev. Med. Child Neurol. 31:640 649.

Cooper, G.P., J.G. Suszkiw, and R.S. Manalis. 1984. Heavy metals: Effects on synaptic transmission. NeuroToxicology 5:247 266.

Cooper, W.C., O. Wong, and L. Kheifets. 1985. Mortality among employees of lead battery plants and lead-producing plants, 1947 - 1980. Scand. J. Work Environ. Health 11:331 345.

Cosgrove, E.V., M.J. Brown, P. Madigan, P. McNulty, L. Okonski, and J. Schmidt. 1989. Childhood lead poisoning: Case study traces source to drinking water. J. Environ. Health 52:346 349.

Crosby, W.H. 1977. Lead-contaminated health food: Association with lead poisoning and leukemia. J. Am. Med. Assoc. 237:2627 2629.

Cumings, J.N. 1959. Part 3: Lead. Pp. 93 124 in Heavy Metals and the Brain. Springfield, Ill.: Charles C Thomas.

Dalgarno, B.G., R.M. Brown, and C.J. Pickford. 1988. Potential of inductively coupled plasma mass spectrometry for trace element metabolism studies in man. Biomed. Environ. Mass Spectrom. 16:377 380.

Date, A.R., and Y.Y. Cheung. 1987. Studies in the determination of lead isotope ratios by inductively coupled plasma mass spectrometry. Analyst 112:1531 1540.

David, O.J., H.L. Wintrob, and C.G. Arcoleo. 1982. Blood lead stability. Arch. Environ. Health 37:147 150.

Davis, A.O., J.N. Galloway, and D.K. Nordstrom. 1982. Lake acidification: Its effect on lead in the sediment of two Adirondack USA lakes. Limnol. Oceanogr. 27:163 167.

Davis, J.M. 1990. Risk assessment of the developmental neurotoxicity of lead. NeuroToxicology 11:285 292.

Davis, J.M., and D.J. Svendsgaard. 1987. Low-level lead exposure and child development. Nature 329:297 300.

Davis, S., P. Waller, R. Buschbom, J. Ballou, and P. White. 1990. Quantitative estimates of soil ingestion in normal children between the ages of 2 and 7 years: Populations-based estimates using aluminum, silicon, and titanium as soil tracer elements. Arch. Environ. Health 45:112 122.

Day, J.P., J.E. Fergusson, and T.M. Chee. 1979. Solubility and potential toxicity of lead in urban street dust. Bull. Environ. Contam. Toxicol. 23:497 502.

Dean, J.R., L. Ebdon, and R. Massey. 1987. Selection of mode for the measurement of lead isotope ratios by inductively coupled plasma mass spectrometry and its application to milk powder analysis. J. Anal. At. Spectrom. 2:369 374.

Dean, J.R., L. Ebdon, H.M. Crews, and R.C. Massey. 1988. Characteristics of flow injection inductively coupled plasma mass spectrometry for trace metal determinations. J. Anal. At. Spectrom. 3:349 354.

de Kort, W.L.A.M., and W.C.M. Zwennis. 1988. Blood lead and blood pressure: Some implications for the situation in the Netherlands. Environ. Health Perspect. 78:67 70.

de Kort, W.L.A.M., M.A. Verschoor, A.A.E. Wibowo, and J.J. van Hemmen. 1987. Occupational exposure to lead and blood pressure: A study in 105 workers. Am. J. Ind. Med. 11:145 156.

Delves, H.T. 1977. Analytical techniques for blood-lead measurements. J. Anal. Toxicol. 1:261 264.
Delves, H.T. 1984. Use reference samples rather than reference methods. Anal. Proc. 21:391 394.
Delves, H.T. 1988. Biomedical applications of ICP-MS. Chem. Br. 24:1009 1012.
Delves, H.T. 1991. Review of the accuracy and precision of methods currently available for blood lead screening. Pp. 117 127 in Proceedings of the First National Conference on Laboratory Issues in Childhood Lead Poisoning Prevention, October 31 November 2, 1991. Washington, D.C.: Association of State and Territorial Public Health Laboratory Directors.
Delves, H.T., and M.J. Campbell. 1988. Measurements of total lead concentrations and of lead isotope ratios in whole blood by use of inductively coupled plasma source mass spectrometry. J. Anal. At. Spectrom. 3:343 348.
Delves, H.T., B.E. Clayton, A. Carmichael, M. Bubear, and M. Smith. 1982. An appraisal of the analytical significance of tooth-lead measurements as possible indices of environmental exposure of children to lead. Ann. Clin. Biochem. 19:329 337.
Delves, H.T., J.C. Sherlock, and M.J. Quinn. 1984. Temporal stability of blood lead concentrations in adults exposed only to environmental lead. Hum. Toxicol. 3:279 288.
DeSilva, P.E. 1981. Determination of lead in plasma and studies on its relationship to lead in erythrocytes. Br. J. Ind. Med. 38:209 217.
Dieter, M.P., and M.T. Finley. 1979. δ-Aminolevulinic acid dehydratase enzyme activity in blood, brain, and liver of lead-dosed ducks. Environ. Res. 19:127 135.
Dietrich, K.N. 1991. Human fetal lead exposure, intra-uterine growth, maturation, and postnatal neurobehavioral development. Fundam. Appl. Toxicol. 16:17 19.
Dietrich, K.N. 1992. Lead Exposure and Central Auditory Processing in Children. Paper presented at the Annual Meeting of the Neurobehavioral Teratology Society, June 1992, Boca Raton, Fla.
Dietrich, K.N., K.M. Krafft, D.T. Pearson, L.C. Harris, R.L. Bornschein, P.B. Hammond, and P.A. Succop. 1985. Contribution of social and developmental factors to lead exposure during the first year of life. Pediatrics 75:1114 1119.

Dietrich K.N., K.M. Krafft, R.L. Bornschein, P.B. Hammond, O. Berger, P.A. Succop, and M. Bier. 1987a. Low-level fetal lead exposure effect on neurobehavioral development in early infancy. Pediatrics 80:721 730.

Dietrich, K.N., K.M. Krafft, R. Shukla, R.L. Bornschein, and P.A. Succop. 1987b. The neurobehavioral effects of early lead exposure. Pp. 71 95 in Toxic Substances and Mental Retardation: Neurobehavioral Toxicology and Teratology. Monographs of the American Association on Mental Deficiency, No. 8, S.R. Schroeder, ed. Washington, D.C.: American Association on Mental Deficiency.

Dietrich, K.N., K.M. Krafft, M. Bier, O. Berger, P.A. Succop, and R.L. Bornschein. 1989. Neurobehavioural effects of fetal lead exposure: The first year of life. Pp. 320 331 in Lead Exposure and Child Development: An International Assessment, M.A. Smith, L.D. Grant, and A.I. Sors, eds. Dordrecht, The Netherlands: Kluwer Academic.

Dietrich, K.N., P.A. Succop, R.L. Bornschein, K.M. Krafft, O. Berger, P.B. Hammond, and C.R. Buncher. 1990. Lead exposure and neurobehavioral development in later infancy. Environ. Health Perspect. 89:13 19.

Dietrich, K.N., P.A. Succop, O.G. Berger, P.B. Hammond, and R.L. Bornschein. 1991. Lead exposure and the cognitive development of urban preschool children: The Cincinnati lead study cohort at age 4 years. Neurotoxicol. Teratol. 13:203 212.

Dietrich, K.N., P.A. Succop, O.G. Berger, and R. Keith. 1992. Lead exposure and the central auditory processing abilities and cognitive development of urban children: The Cincinnati lead study cohort at age 5 years. Neurotoxicol. Teratol. 14:51 56.

Dietrich, K., O. Berger, P. Succop, P. Hammond, and R. Bornschein. 1993a. The developmental consequences of low to moderate prenatal and postnatal lead exposure: Intellectual attainment in the Cincinnati Lead Study Cohort following school entry. Neurotoxicol. Teratol. 15:37 44.

Dietrich, K., O. Berger, and P. Succop. 1993b. Lead exposure and the motor developmental status of urban 6-year-old children in the Cincinnati Prospective Study. Pediatrics 91:301 307.

Diver, F.S., D.L. Littlejohn, T.D.B. Lyon, and G.S. Fell. 1988.

Human albumin as a reference material for trace elements. Fresenius Z. Anal. Chem. 332:627 629.

Doherty, W. 1989. An internal standardization procedure for the determination of yttrium and the rare earth elements in geological materials by inductively coupled plasma-mass spectrometry. Spectrochim. Acta 44B(3):263 280.

Dornan, J. 1986. Lead absorption in the mineral extraction industry. J. Soc. Occup. Med. 36:99 101.

Doss, M., and W.A. Muller. 1982. Acute lead poisoning in inherited porphobilinogen synthase (δ-aminolevulinic acid dehydrase) deficiency. Blut 45:131 139.

Doss, M., R. von Tiepermann, J. Schneider, and H. Schmid. 1979. New type of hepatic porphyria with porphobilinogen synthase defect and intermittent acute clinical manifestation. Klin. Wochenschr. 57:1123 1127.

Doss, M., U. Becker, F. Sixel, S. Geisse, H. Solcher, and J. Schneider. 1982. Persistent protoporphyrinemia in hereditary porphobilinogen synthase (δ-aminolevulinic acid dehydrase) deficiency under low lead exposure: A new molecular basis for the pathogenesis of lead intoxication. Klin. Wochenschr. 60:599 606.

Douglas, D.J., and R.S. Houk. 1985. Inductively-coupled plasma mass spectrometry (ICP-MS). Prog. Anal. At. Spectrosc. 8:1 18.

Douglas, D.J., E.S.K. Quan, and R.G. Smith. 1983. Elemental analyses with an atmospheric pressure plasma (MIP, ICP)/quadrupole mass spectrometer system. Spectrochim. Acta 38B:39 48.

Drasch, G.A., and J. Ott. 1988. Lead in human bones. Investigations on an occupationally non-exposed population in southern Bavaria (F.R.G.): II. Children. Sci. Total Environ. 68:61 69.

Drasch, G.A., J. Bohm, and C. Bauer. 1987. Lead in human bones. Investigations on an occupationally non-exposed population in southern Bavaria (F.R.G.): I. Adults. Sci. Total Environ. 64:303 315.

Duggan, M.J. 1983. The uptake and excretion of lead by young children. Arch. Environ. Health 38:246 247.

Duggan, M.J., and M.J. Inskip. 1985. Childhood exposure to lead in surface dust and soil: A community health problem. Public Health Rev. 13:1 54.

DuVal, G., and B.A. Fowler. 1989. Preliminary purification and characterization studies of a low molecular weight, high affinity cytosolic lead-binding protein in rat brain. Biochem. Biophys. Res. Commun. 159:177 184.

DuVal, G.E., D.A. Jett, and B.A. Fowler. 1989. Lead-induced aggregation of α_2u-globulin in vitro [abstract]. Toxicologist 9:98.

Dzubay, T.G., R.K. Stevens, and L.W. Richards. 1979. Composition of aerosols over Los Angeles freeways. Atmos. Environ. 13:653 659.

Edelman, G.M. 1986. Cell adhesion molecules in the regulation of animal form and tissue pattern. Annu. Rev. Cell Biol. 2:81 116.

Edelstein, S., C.S. Fullmer, and R.H. Wasserman. 1984. Gastrointestinal absorption of lead in chicks: Involvement of the cholecalciferol endocrine system. J. Nutr. 114:692 700.

Edgington, D.N., and J.A. Robbins. 1976. Records of lead deposition in Lake Michigan sediments since 1800. Environ. Sci. Technol. 10:266 274.

Ediger, R.D., and R.L. Coleman. 1972. A modified Delves cup atomic absorption procedure for the determination of lead in blood. Atomic Absorp. Newsl. 11:33 36.

Egeland, G.M. G.A. Burkhart, T.M. Schnorr, R.W. Hornung, J.M. Fajen, and S.T. Lee. 1992. Effects of exposure to carbon disulphide on low-density lipoprotein cholesterol concentration and diastolic blood pressure. Br. J. Ind. Med. 49:287 293.

Elbaz-Poulichet, F., P. Holliger, W.W. Huang, and J.-M. Martin. 1984. Lead cycling in estuaries, illustrated by the Gironde estuary, France. Nature 308:409 414.

Elias, R.W., Y. Hirao, and C.C. Patterson. 1982. The circumvention of the natural biopurification of calcium along nutrient pathways by atmospheric inputs of industrial lead. Geochim. Cosmochim. Acta 46:2561 2580.

Elinder, C.G., L. Friberg, B. Lind, B. Nilsson, M. Svartengren, and I. Overmark. 1986. Decreased blood lead levels in residents of Stockholm for the period 1980 1984. Scand. J. Work Environ. Health 12:114 120.

Elinder, C.G., B. Lind, B. Nilsson, and A. Oskarsson. 1988. Wine—An important source of lead exposure. Food Addit. Contam. 5:641 644.

Elwood, P., and C. Toothill. 1986. Further evidence of a fall in blood lead levels in Wales. J. Epidemiol. Commun. Health 40:178 180.

Elwood, P.C., G. Davey-Smith, P.D. Oldham, and C. Toothill. 1988a. Two Welsh surveys of blood lead and blood pressure. Environ. Health Perspect. 78:119 121.

Elwood, P.C., J.W.G. Yarnell, P.D. Oldham, J.C. Catford, D. Nutbeam, G. Davey-Smith, and C. Toothill. 1988b. Blood pressure and blood lead in surveys in Wales. Am. J. Epidemiol. 127:942 45.

Emmerson, B.T. 1963. Chronic lead nephropathy: The diagnostic use of calcium EDTA and the association with gout. Aust. Ann. Med. 12:310 324.

Emmerson, B.T. 1973. Chronic lead nephropathy [editorial]. Kidney Int. 4:1 5.

Emmerson, B.T., and D.S. Lecky. 1963. The lead content of bone in subjects without recognized past lead exposure and in patients with renal disease. Aust. Ann. Med. 12:139 142.

Englert, N. 1978. Measurement of peripheral motor nervous conduction velocity in adults and children with elevated lead blood level [in German]. BGA Ber., No. 1:108 117.

EPA (U.S. Environmental Protection Agency). 1973. Regulation of fuels and fuel additives: Control of lead additives in gasoline. Fed. Regist. 38(234):33734 33741.

EPA (U.S. Environmental Protection Agency). 1982. Regulation of fuels and fuel additives. Fed. Regist. 47(210):49322 49334.

EPA (U.S. Environmental Protection Agency). 1985. Regulations of Fuels and Fuel Additives: Gasoline Lead Content. Fed. Regist. 50(45):9386 9408.

EPA (U.S. Environmental Protection Agency). 1986a. Air Quality Criteria for Lead. EPA-600/08-83/028aF-dF. Research Triangle Park, N.C.: U.S. Environmental Protection Agency. 4 Vols.

EPA (U.S. Environmental Protection Agency). 1986b. Guidelines for carcinogen risk assessment. Fed. Regist. 51(185):33992 34003.

EPA (U.S. Environmental Protection Agency). 1987. Preliminary Results: The Aging Solder Study. Office of Drinking Water. Washington, D.C.: U.S. Environmental Protection Agency.

EPA (U.S. Environmental Protection Agency). 1990a. Report of the Clean Air Science Advisory Committee (CASAC): Review of the OAQPS Lead Staff Paper and the ECAO Air Quality Criteria Document Supplement. EPA-SAB-CASAC-90-002. Washington, D.C.: U.S. Environmental Protection Agency.

EPA (U.S. Environmental Protection Agency). 1990b. Air Quality Criteria for Lead: Supplement to the 1986 Addendum. EPA-600/08-89/049f. Research Triangle Park, N.C.: U.S. Environmental Protection Agency.

EPA (U.S. Environmental Protection Agency). 1990c. Review of the National Ambient Air Quality Standards for Lead: Assessment of Scientific and Technical Information. EPA-450/02-89/022. Office of Air Quality Planning and Standards. Research Triangle Park, N.C.: U.S. Environmental Protection Agency.

EPA (U.S. Environmental Protection Agency). 1991. Maximum contaminant level goals and national primary drinking water regulations for lead and copper. Fed. Regist. 56(110):26460 26564.

Ericson, J.E., H. Shirahata, and C.C. Patterson. 1979. Skeletal concentrations of lead in ancient Peruvians. N. Engl. J. Med. 300: 946 951.

Ericson, J.E., D.R. Smith, and A.R. Flegal. 1991. Skeletal concentrations of lead, cadmium, zinc, and silver in ancient North American Pacos Indians. Environ. Health Perspect. 93:217 223.

Erkkilä, J., R. Armstrong, V. Riihimäki, D.R. Chettle, A. Paakkari, M. Scott, L. Somervaille, J. Starck, B. Kock, and A. Aitio. 1992. In vivo measurements of lead in bone at four anatomical sites: Long-term occupational and consequent endogenous exposure. Br. J. Ind. Med. 49:631 644.

Ernhart, C.B., and T. Greene. 1990. Low-level lead exposure in the prenatal and early preschool periods: Language development. Arch. Environ. Health 45:342 354.

Ernhart, C.B., and M. Morrow-Tlucak. 1987. Low level lead exposure in the prenatal and early preschool years as related to intelligence just prior to school entry. Pp. 150 152 in International Conference on Heavy Metals in the Environment, Vol. 1, S.E. Lindberg and T.C. Hutchinson, eds. Edinburgh: CEP Consultants.

Ernhart, C., B. Landa, and N.B. Schell. 1981. Subclinical levels of lead and developmental deficit: A multivariate follow-up reassessment. Pediatrics 67:911 919.

Ernhart, C.B., A.W. Wolf, P.L. Linn, R.J. Sokol, M. Kennard, and H. Filipovich. 1985. Alcohol-related birth defects: Syndromal anomalies, intrauterine growth retardation, and neonatal behavioral assessment. Alcohol Clin. Exp. Res. 9:447 453.

Ernhart, C.B., A.W. Wolf, M.J. Kennard, P. Erhard, H.F. Filipovich, and R.J. Sokol. 1986. Intrauterine exposure to low levels of lead: The status of the neonate. Arch. Environ. Health 41:287 291.

Ernhart, C.B., M. Morrow-Tlucak, M.R. Marler, and A.W. Wolf. 1987. Low level lead exposure in the prenatal and early preschool periods: Early preschool development. Neurotoxicol. Teratol. 9:259 270.

Ernhart, C.B., M. Morrow-Tlucak, A.W. Wolf, D. Super, and D. Drotar. 1989. Low level lead exposure in the prenatal and early preschool periods: Intelligence prior to school entry. Neurotoxicol. Teratol. 11:161 170.

Everson, J., and C.C. Patterson. 1980. "Ultra-clean" isotope dilution/mass spectrometric analyses for lead in human blood plasma indicate that most reported values are artificially high. Clin. Chem. 26:1603 1607.

Evis, M.J., K. Dhaliwal, K.A. Kane, M.R. Moore, and J.R. Parratt. 1987. The effects of chronic lead treatment and hypertension on the severity of cardiac arrhythmias induced by coronary artery occlusion or by noradrenalin in anesthetised rats. Arch. Toxicol. 59:336 340.

Ewers, U., D. Kello, U. Krämer, M. Neuf, and G. Winneke. 1989. Formal Report of the Joint WHO/CEC Final Review Workshop on the Lead Neurotoxicity Study in Children. May 9 12, 1989. Düsseldorf, Germany.

Facchetti, S., and F. Geiss. 1982. Isotopic Lead Experiment: Status Report, Publ. No. EUR 8352 EN. Luxembourg: Commission of the European Communities.

Factor-Litvak, P., J.H. Graziano, J.K. Kline, D. Popovac, A. Mehmeti, G. Ahmedi, P. Shrout, M.J. Murphy, E. Gashi, R. Haxhiu, L. Rajovic, D.U. Nenezic, and Z.A. Stein. 1991. A prospective study of birthweight and length of gestation in a population surrounding a lead smelter in Kosovo, Yugoslavia. Int. J. Epidemiol. 20: 722 728.

Fahim, M.S., Z. Fahim, and D.G. Hall. 1976. Effects of subtoxic lead levels on pregnant women in the state of Missouri. Res. Commun. Chem. Pathol. Pharmacol. 13:309 331.

Fanning, D. 1988. A mortality study of lead workers, 1926 1985. Arch. Environ. Health 43:247 251.

Farfel, M.R. 1985. Reducing lead exposure in children. Annu. Rev. Public Health 6:333 360.

Farfel, M.R. 1987. Evaluation of health and environmental effects of two methods for residential lead paint removal [dissertation]. Baltimore, Md.: Johns Hopkins University.

Farfel, M.R., and J.J. Chisolm, Jr. 1987. Comparison of traditional and alternative residential lead paint removal methods. Pp. 212 214 in International Conference on Heavy Metals in the Environment, Vol. 2, S.E. Lindberg and T.C. Hutchinson, eds. Edinburgh: CEP Consultants.

Farkas, W.R., A. Fischbein, S. Solomon, F. Buschman, E. Borek, and O.K. Sharma. 1987. Elevated urinary excretion of β-aminoisobutyric acid and exposure to inorganic lead. Arch. Environ. Health 42:96 99.

Faure, G. 1986. Principles of Isotope Geology, 2nd Ed. New York: John Wiley & Sons. 589 pp.

Feldman, R.G., J. Haddow, and J.J. Chisolm, Jr. 1973a. Chronic lead intoxication in urban children: Motor nerve conduction velocity studies. Pp. 313 317 in New Developments in Electromyography and Clinical Neurophysiology, Vol. 2, J. Desmedt and S. Karger, eds. Basel, Switzerland: Karger.

Feldman, R.G., J. Haddow, L. Kopito, and H. Schwachman. 1973b. Altered peripheral nerve conduction velocity: Chronic lead intoxication in children. Am. J. Dis. Child. 125:39 41.

Feldman, R.G., M.K. Hayes, R. Younes and F.D. Aldrich. 1977. Lead neuropathy in adults and children. Arch. Neurol. 34:481 488.

Fergusson, J.E., and N.G. Purchase. 1987. The analysis and levels of lead in human teeth: A review. Environ. Pollut. 46:11 44.

Fergusson, J.E., and R.J. Schroeder. 1985. Lead in house dust of Christchurch, New Zealand: Sampling, levels and sources. Sci. Total Environ. 46:61 72.

Fergusson, D.M., J.E. Fergusson, L.J. Horwood, and N.G. Kinzett. 1988. A longitudinal study of dentine lead levels, intelligence, school performance and behaviour: Part II. Dentine lead and cognitive ability. J. Child Psychol. Psychiatr. 29:793 809.

Finelli, V.N., D.S. Klauder, M.A. Karaffa, and H.G. Petering. 1975. Interaction of zinc and lead on δ-aminolevulinate dehydrase. Biochem. Biophys. Res. Commun. 65:303 311.

Fison, D.C. 1978. The Royal Children's Hospital, Brisbane: 1878 to 1978 [editorial]. Med. J. Aust. 2:137 138.

Flanagan, P.R., M.J. Chamberlain, and L.S. Valberg. 1982. The relationship between iron and lead absorption in humans. Am. J. Clin. Nutr. 36:823 829.

Flegal, A.R. 1986. Lead in tropical marine systems: A review. Sci. Total Environ. 58:1 8.

Flegal, A.R., and K.H. Coale. 1989. Comments on "Trends in lead concentrations in major U.S. rivers and their relation to historical changes in gasoline-lead consumption." Water Resour. Bull. 25: 1275 1277.

Flegal, A.R., and C.C. Patterson. 1983. Vertical concentration profiles of lead in the Central Pacific at 15°N and 20°S. Earth Planet. Sci. Lett. 64:19 32.

Flegal, A.R., and D.R. Smith. 1992. Current needs for increased accuracy and precision in measurements of low levels of lead in blood. Environ. Res. 58:125 133.

Flegal, A.R., and V.J. Stukas. 1987. Accuracy and precision of lead isotopic composition measurements in seawater. Marine Chem. 22:163 177.

Flegal, A.R., K.J. Rosman, and M.D. Stephenson. 1987. Isotope systematics of contaminant leads in Monterey Bay. Environ. Sci. Technol. 21:1075 1079.

Flegal, A.R., J.O. Nriagu, S. Niemeyer, and K.H. Coale. 1989. Isotopic tracers of lead contamination in the Great Lakes. Nature 339:455 458.

Foreman, H. 1963. Toxic side effects of ethylenediaminetetraacetic acid. J. Chron. Dis. 16:319 323.

Fowler, B.A., and G. DuVal. 1991. Effects of lead on the kidney: Roles of high-affinity lead-binding proteins. Environ. Health Perspect. 91:77 80.

Fowler, B.A., C.A. Kimmel, J.S. Woods, E.E. McConnell, and L.D. Grant. 1980. Chronic low level lead toxicity in the rat: III. An integrated assessment of long-term toxicity with special reference to the kidney. Toxicol. Appl. Pharmacol. 56:59 77.

Fowler, B.A., P. Mistry, and W.W. Victery. 1985. Ultrastructural morphometric studies of lead intranuclear inclusion body formation in kidney proximal tubule cells: Relationship to altered renel protein synthetic patterns [abstract]. Toxicologist 5:53.

Friberg, L.T. 1985. Yant memorial lecture. The rationale of biological monitoring of chemicals With special reference to metals. Am. Ind. Hyg. Assoc. J. 46:633 642.

Friberg, L. 1988. Quality assurance. Pp. 103 126 in Biological Monitoring of Toxic Metals, Rochester Series on Environmental Toxicity, T.W. Clarkson, L. Friberg, G.F. Nordberg, and P.R. Sager, eds. New York: Plenum.

Friedlander, S.K. 1977. Introduction: Aerosol characterization. Pp. 1 23 in Smoke, Dust and Haze: Fundamentals of Aerosol Behavior. New York: John Wiley & Sons.

Friedstein, H.G. 1981. A short history of the chemistry of painting. J. Chem. Educ. 58:291 295.

Frisancho, A., and A. Ryan. 1991. Decreased stature associated with moderate blood lead concentrations in Mexican-American children. Am. J. Clin. Nutr. 54:516 519.

Fullerton, P.M. 1966. Chronic peripheral neuropathy produced by lead poisoning in guinea pigs. J. Neuropathol. Exp. Neurol. 25:214 236.

Fullmer, C.S., S. Edelstein, and R.H. Wasserman. 1985. Lead-binding properties of intestinal calcium-binding proteins. J. Biol. Chem. 260:6816 6819.

Fulton, M., G. Raab, G. Thomson, D. Laxen, R. Hunter, and W. Hepburn. 1987. Influence of blood lead on the ability and attainment of children in Edinburgh. Lancet 1:1221 1226.

Galloway, J.N., J.D. Thornton, S.A. Norton, H.L. Volchok, and R.A. McLean. 1982. Trace metals in atmospheric deposition. A review and assessment. Atmos. Environ. 16:1677 1700.

Gardels, M.C. 1989. Laboratory study of lead leaching from drinking water coolers [abstract]. Am. Water Works Assoc. J. 81:73.

Gardels, M.C. and T.J. Sorg. 1989. A laboratory study of the leaching of lead from water faucets. J. Am. Water Works Assoc. 81: 101 113.

Gebhart, A.M. and G.W. Goldstein. 1988. Use of an in vitro system to study the effects of lead on astrocyte-endothelial cell interactions: A model for studying toxic injury to the blood-brain barrier. Toxicol. Appl. Pharmacol. 94:191 206.

Gerhardsson, L., N.-G. Lundstrom, G. Nordberg, and S. Wall. 1986. Mortality and lead exposure: A retrospective cohort study of Swedish smelter workers. Br. J. Ind. Med. 43:707 712.

Gibson, J.L. 1904. A plea for painted railings and painted walls of rooms as the source of lead poisoning among Queensland children. Aust. Med. Gazette 23:149 53.
Gibson, J.L, W. Love, D. Hardine, P. Bancroft, and A.J. Turner. 1892. Note on lead poisoning as observed among children in Brisbane. Pp. 76 83 in Transactions of the Third Intercolonial Medical Congress of Australasia, L.R. Huxtable, ed. Sydney, Australia: Charles Potter.
Gilfillan, S.C. 1965. Lead poisoning and the fall of Rome. J. Occup. Med. 7:53 60.
Goering, P.L. In press. Lead-protein interactions as a basis for lead toxicity. NeuroToxicology 41(2).
Goering, P.L., and B.A. Fowler. 1984. Regulation of lead inhibition of δ-aminolevulinic acid dehydratase by a low molecular weight, high affinity renal lead-binding protein. J. Pharmacol. Exp. Ther. 231:66 71.
Goering, P.L., and B.A. Fowler. 1985. Mechanism of renal lead-binding protein reversal of δ-aminolevulinic acid dehydratase inhibition by lead. J. Pharmacol. Exp. Ther. 234:365 371.
Goering, P.L., P. Mistry, and B.A. Fowler. 1986. A low molecular weight lead-binding protein in brain attenuates lead inhibition of δ-aminolevulinic acid dehydratase: Comparison with a renal lead-binding protein. J. Pharmacol. Exp. Ther. 237:220 225.
Goldberg, A., and M.R. Moore, eds. 1980. The Porphyrias. Clinics in Haematology, Vol. 9. London: Saunders.
Goldstein, G.W. In press. Evidence that lead acts as a calcium substitute in second messenger metabolism. NeuroToxicology
Goyer, R.A. 1989. Mechanisms of lead and cadmium nephrotoxicity. Toxicol. Lett. 46:153 162.
Goyer, R.A., and B.C. Rhyne. 1973. Pathological effects of lead. Int. Rev. Exp. Pathol. 12:1 77.
Graham, D.L. and S.M. Kalman. 1974. Lead in forage grass from a suburban area in northern California. Environ. Pollut. 7:209 215.
Grandjean, P. 1978. Regional distribution of lead in human brains. Toxicol. Lett. 2:65 69.
Grandjean, P., and J. Lintrup. 1978. Erythrocyte Zn-protoporphyrin as an indicator of lead exposure. Scand. J. Clin. Lab. Invest. 38: 669 675.
Grandjean, P., O.N. Hansen, and T. Lyngbye. 1984. Analysis of lead

in circumpulpal dentin of deciduous teeth. Ann. Clin. Lab. Sci. 14:270 275.

Grandjean, P., T. Lyngbye, and O. Hansen. 1986. Lead concentration in deciduous teeth: Variation related to tooth type and analytical technique. J. Toxicol. Environ. Health 19:437 445.

Grandjean, P., H. Hollnagel, L. Hedegaard, J.M. Christensen, and S. Larsen. 1989. Blood lead-blood pressure relations: Alcohol intake and hemoglobin as confounders. Am. J. Epidemiol. 129:732 734.

Granick, J.L., S. Sassa, S. Granick, R.D. Levere, and A. Kappas. 1973. Studies in lead poisoning: II. Correlation between the ratio of activated to inactivated δ-aminolevulinic acid dehydratase of whole blood and the blood lead level. Biochem. Med. 8:149 159.

Granick, J.L., S. Sassa, and A. Kappas. 1978. Some biochemical and clinical aspects of lead intoxication. Adv. Clin. Chem. 20:288 339.

Grant, L.D., and J.M. Davis. 1989. Effects of low-level lead exposure on paediatric neurobehavioural development: Current findings and future directions. Pp. 49 115 in Lead Exposure and Child Development: An International Assessment, M.A. Smith, L.D. Grant, and A.I. Sors, eds. Dordrecht, The Netherlands: Kluwer Academic.

Gray, A.L. 1989. The origins, realization and performance of ICP-MS systems. Pp. 1 42 in Applications of Inductively Coupled Plasma Mass Spectrometry, A.R. Date and A.L. Gray, eds. New York: Chapman and Hall.

Graziano, J., D. Popovac, M. Murphy, A. Mehmeti, J. Kline, G. Ahmedi, P. Shrout, Z. Zvicer, G. Wasserman, E. Gashi, Z. Stein, B. Rajovic, L. Belmont, B. Colakovic, R. Bozovic, R. Haxhiu, L. Radovic, R. Vlaskovic, D. Nenezic, and N. Lolacono. 1989. Environmental lead, reproduction and infant development. Pp. 379 386 in Lead Exposure and Child Development: An International Assessment, M.A. Smith, L.D. Grant, and A.I. Sors, eds. Dordrecht, The Netherlands: Kluwer Academic.

Graziano, J.H., D. Popovac, P. Factor-Litvak, P. Shrout, J. Kline, M.J. Murphy, Y.-h. Zhao, A. Mehmeti, X. Ahmedi, G. Ahmedi, B. Rajovic, Z. Zvicer, D.U. Nenezic, N.J. Lolacono, and Z. Stein. 1990. Determinants of elevated blood lead during pregnancy in a population surrounding a lead smelter in Kosovo, Yugoslavia. Environ. Health Perspect. 89:95 100.

Green, S., D.A. Bradley, J.E. Palethorpe, D. Mearman, D.R. Chettle, A.D. Lewis, P.J. Mountford, and W.D. Morgan. 1993. An en-

hanced sensitivity K-shell x-ray fluorescence technique for tibial lead determination. Phys. Med. Biol. 38:389 396.
Greene, T., and C.B. Ernhart. 1991. Prenatal and preschool age lead exposure: Relationship with size. Neurotoxicol. Teratol. 13:417 427.
Griffin, T.B., F. Coulston, L. Goldberg, H. Wills, J.C. Russell, and J.H. Knelson. 1975. Clinical studies on men continuously exposed to airborne particulate lead. Pp. 221 240 in Lead, Vol. 2, Supplement, T.B. Griffin, and J.H. Knelson, eds. Stuttgart, Germany: Georg Thieme Publishers.
Griliches, Z. 1977. Estimating the returns to schooling: Some econometric problems. Econometrica 45:1 22.
Gross, S.B., E.A. Pfitzer, D.W. Yeager, and R.A. Kehoe. 1975. Lead in human tissues. Toxicol. Appl. Pharmacol. 32:638 651.
Gualtieri, C.T. 1987. Fetal antigenicity and maternal immunoreactivity: Factors in mental retardation. Pp. 33 69 in Toxic Substances and Mental Retardation: Neurobehavioral Toxicology and Teratology. Monographs of the American Association on Mental Deficiency, No. 8, S.R. Schroeder, ed. Washington, D.C.: American Association on Mental Deficiency.
Habermann, E., K. Crowell, and P. Janicki. 1983. Lead and other metals can substitute for Ca^{2+} in calmodulin. Arch. Toxicol. 54:61 70.
Hachey, D.L., W.W. Wong, T.W. Boutton, and P.D. Klein. 1987. Isotope ratio measurements in nutrition and biomedical research. Mass Spectrom. Rev. 6:289 328.
Hall, A., and M.D. Cantab. 1905. The increasing use of lead as an abortifacient: A series of thirty cases of plumbism. Br. Med. J. 1:584 587.
Hamelin, B., B. Dupre, O. Brevart, and C.J. Allegre. 1988. Metallogenesis at paleo-spreading centers: Lead isotopes in sulfides, rocks, and sediments from the *Troodos ophiolite (Cyprus)*. Chem. Geol. 68:229 238.
Hamilton, A., and H.L. Hardy. 1949. Industrial Toxicology, 2nd Ed. New York: Hoeber.
Hammond, P.B., R.L. Bornschein, and P. Succop. 1985. Dose-effect and dose-response relationships of blood lead to erythrocytic protoporphyrin in young children. Environ. Res. 38:187 196.
Hansen, O.N., A. Trillingsgaard, I. Beese, T. Lyngbye, and P. Grand-

jean. 1989a. Neuropsychological profile of children in relation to dentine lead level and socioeconomic group. Pp. 240 250 in Lead Exposure and Child Development: An International Assessment, M.A. Smith, L.D. Grant, and A.I. Sors, eds. Dordrecht, The Netherlands: Kluwer Academic.

Hansen, O.N., A. Trillingsgaard, I. Beese, T. Lyngbye, and P. Grandjean. 1989b. A neuropsychological study of children with elevated dentine lead level: Assessment of the effect of lead in different socio-economic groups. Neurotoxicol. Teratol. 11:205 213.

Harlan, W.R., J.R. Landis, R.L. Schmouder, N.G. Goldstein, and L.C. Harlan. 1985. Blood lead and blood pressure: Relationship in the adolescent and adult US population. J. Am. Med. Assoc. 253: 530 534.

Harley, N.H., and T.H. Kneip. 1984. An Integrated Metabolic Model for Lead in Humans of All Ages. Final Report to the U.S. Environmental Protection Agency, Contract No. B44899 with New York University School of Medicine. Washington, D.C.: U.S. Environmental Protection Agency.

Harris, R.W., and W.R. Elsea. 1967. Ceramic glaze as a source of lead poisoning. J. Am. Med. Assoc. 202:544 546.

Harrison, R.M., and D.P.H. Laxen. 1981. Measurements of gaseous lead alkyls in polluted atmospheres. Atmos. Environ. 15:422 424.

Harrison, G.E., T.E.F. Carr, A. Sutton, E.R. Humphreys, and Rundo. 1969. Effect of alginate on the absorption of lead in man. Nature 224:1115 1116.

Harrison, R.M., and C.R. Williams. 1982. Airborne cadmium, lead and zinc at rural and urban sites in north-west England. Atmos. Environ. 16:2669 2681.

Harrison, W.W. 1988. Glow discharge mass spectrometry: A current assessment. J. Anal. At. Spectrom. 3:867 872.

Harrison, W.W., and B.L. Bentz. 1988. Glow discharge mass spectrometry. Prog. Anal. Spectrosc. 11:53 110.

Harvey, P.G., M.W. Hamlin, R. Kumar, and H.T. Delves. 1984. Blood lead, behavior, and intelligence test performance in pre-school children. Sci. Total Environ. 40:45 60.

Hatzakis, A., A. Kokkevi, K. Katsouvanni, K. Maravelias, F. Salaminios, A. Kalandidi, A. Koutselinis, K. Stefanis, and D. Trichopoulos. 1987. Psychometric intelligence and attentional performance deficits in lead-exposed children. Pp. 204 209 in International Conference on Heavy Metals in the Environment, Vol. 1, S.E. Lindberg and T.C. Hutchinson, eds. Edinburgh: CEP Consultants.

Hatzakis, A., A. Kokkevi, C. Maravelias, K. Katsouyanni, F. Salaminios, A. Kalandidi, A. Koutselinis, C. Stefanis, and D. Trichopoulos. 1989. Psychometric intelligence deficits in lead-exposed children. Pp. 211 223 in Lead Exposure and Child Development: An International Assessment, M.A. Smith, L.D. Grant, and A.I. Sors, eds. Dordrecht, The Netherlands: Kluwer Academic.

Hawk, B.A., S.R. Schroeder, G. Robinson, D. Otto, P. Mushak, D. Kleinbaum, and G. Dawson. 1986. Relation of lead and social factors to IQ of low SES children: A partial replication. Am. J. Ment. Defic. 91:178 183.

Healy, M.A., P.G. Harrison, M. Aslem, S.S. Davis, and C.G. Wilson. 1982. Lead sulphide and traditional preparations: Routes for ingestion, solubility, and reactions in gastric fluid. J. Clin. Hosp. Pharm. 7:169 173.

Heard, M.J., and A.C. Chamberlain. 1982. Effect of minerals and food on uptake of lead from the gastrointestinal tract of humans. Hum. Toxicol. 1:411 415.

Heard, M.J., A.C. Chamberlain, and J.C. Sherlock. 1983. Uptake of lead by humans and effect of minerals and food. Sci. Total Environ. 30:245 253.

Henderson, D.A. 1954. A follow-up of cases of plumbism in children. Aust. Ann. Med. 3:219 224.

Herberson, B.M. A.J. King, and J. Allen. 1987. Epithelial cell proliferation in the rat urinary system induced by parenteral injection of lead salts. Br. J. Exp. Pathol. 68:167 177.

Hernberg, S., and J. Nikkanen. 1970. Enzyme inhibition by lead under normal urban conditions. Lancet 1(7637):63 64.

Heumann, K.G. 1988. Isotope dilution mass spectrometry. Pp. 301 376 in Inorganic Mass Spectrometry, F. Adams, R. Gijbels, and R. Van Grieken, eds. New York: John Wiley & Sons.

Heyworth, F., J. Spickett, M. Dick, B. Margetts, and B. Armstrong. 1981. Tailings from a lead mine and lead levels in school children. Med. J. Aust. 2:232 234.

Hiasa, Y., M. Ohshima, Y. Kitahori, T. Fujita, T. Yuasa, and A. Miyashiro. 1983. Basic lead acetate: Promoting effect on the development of renal tubular cell tumors in rats treated with *N*-ethyl-*N*-hydroxyethylnitrosamine. J. Natl. Cancer Inst. 70:761 765.

Hieftje, G.M., and G.H. Vickers. 1989. Developments in plasma source/mass spectrometry. Anal. Chim. Acta 216:1 24.

Hitzfeld, B., F. Planas-Bohne, and D. Taylor. 1989. The effect of lead on protein and DNA metabolism of normal and lead-adapted rat kidney cells in culture. Biol. Trace Elem. Res. 21:87 95.

Hofmann, W. 1982. Dose calculations for the respiratory tract from inhaled natural radioactive nuclides as a function of age: II. Basal cell dose distributions and associated lung cancer risk. Health Phys. 43:31 44.

Hofmann, W., H. Steinhausler, and E. Pohl. 1979. Dose calculation for the respiratory tract from inhaled natural radioactive nuclides as a function of age. I. Compartmental deposition, retention and resulting dose. Health Phys. 37:517 532.

Holtzman, D., J.S. Hsu, and M. Desautel. 1981. Absence of effects of lead feedings and growth-retardation on mitochondrial and microsomal cytochromes in the developing brain. Toxicol. Appl. Pharmacol. 58:48 56.

Horiguchi, S., K. Teramoto, H. Nakano, K. Shinagawa, and G. Endo. 1974. Osmotic fragility test of red blood cells of lead workers by coil planet centrifuge. Osaka City Med. J. 20:51 53.

Houk, R.S., and J.J. Thompson. 1988. Inductively coupled plasma mass spectrometry. Mass Spectrom. Rev. 7:425 461.

Hryhorczuk, D.O., M.B. Rabinowitz, S.M. Hessl, D. Hoffman, M.M. Hogan, K. Mallin, H. French, P. Arris, and E. Berman. 1985. Elimination kinetics of blood lead in workers with chronic lead intoxication. Am. J. Ind. Med. 8:33 42.

Hu, H., F.L. Milder, and D.E. Burger. 1990. X-ray fluorescence measurements of lead burden in subjects with low-level community lead exposure. Arch. Environ. Health 45:335 341.

Hu, H. 1991. A 50-year follow-up of childhood plumbism: Hypertension, renal function, and hemoglobin levels amoung survivors. Am. J. Dis. Child. 145:681 687.

Huang, L.Q., S.-J. Jiang, and R.S. Houk. 1987. Scintillation-type ion detection for inductively coupled plasma mass spectrometry. Anal. Chem. 59:2316 2320.

HUD (U.S. Department of Housing and Urban Development). 1990.

Comprehensive and Workable Plan for the Abatement of Lead-Based Paint in Privately Owned Housing: Report to Congress. Washington, D.C.: U.S. Department of Housing and Urban Development.
Huisman, A., and J.G. Aarnoudse. 1986. Increased 2nd trimester hemoglobin concentration in pregnancies later complicated by hypertension and growth retardation. Acta Obstet. Gynecol. Scand. 65:605 608.
Huneke, J.C. 1988. Relative sensitivity and quantitation in glow-discharge mass-spectrometry A progress report. J. Res. Natl. Bur. Stand. 93:392 393.
Hunt, T.J., R. Hepner, and K.W. Seaton. 1982. Childhood lead poisoning and inadequate child care. Am. J. Dis. Child. 136:538 542.
Hunter, D. 1978. The ancient metals. Pp. 248 297 in The Diseases of Occupations, 6th Ed. London: Hodder & Stoughton.
Hunter, J., M. Urbanowicz, W. Yule, and R. Lansdown. 1985. Automated testing of reaction time and its association with lead in children. Int. Arch. Occup. Environ. Health 57:27 34.
Huntzicker, J.J., S.K. Friedlander, and C.I. Davidson. 1975. Material balance for automobile-emitted lead in Los Angeles basin. Environ. Sci. Technol. 9:448 457.
Hursh, J.B., and J. Suomela. 1968. Absorption of ^{212}Pb from the gastrointestinal tract of man. Acta Radiol. Ther. Phys. Biol. 7:108 120.
Huseman, C.A., C.M. Moriarty, and C.R. Angle. 1987. Childhood lead toxicity and impaired release of thyrotropin-stimulating hormone. Environ. Res. 42:524 533.
Hutton, R.C., and A.N. Eaton. 1988. Analysis of solutions containing high levels of dissolved solids by inductively coupled plasma mass spectrometry. J. Anal. At. Spectrom. 3:547 550.
Iannaccone, A., M. Carmignani, and P. Boscolo. 1981. Cardiovascular reactivity of the rat after chronic exposure to cadmium or lead[in Italian]. Ann. Ist. Super. Sanita 17:655 660.
IARC (International Agency for Research on Cancer). 1980. Lead and lead compounds. Pp. 325 415 in IARC Monographs on the Evaluation of Carcinogenic Risk of Chemicals to Humans. Some Metals and Metalic Compounds, Vol. 23. Lyon, France: International Agency for Research on Cancer.

IARC (International Agency for Research on Cancer). 1987. Overall evaluations of carcinogenicity. Lead and lead compounds. Pp. 230 232 in IARC Monographs on the Evaluation of Carcinogenic Risks to Humans. Overall Evaluations of Carcinogenicity: An Updating of IARC Monographs Vols. 1 42, Suppl. 7. Lyon, France: International Agency for Research on Cancer.

ICRP (International Commission on Radiological Protection). 1991. Quantities used in radiological protection. Pp. 4 11 in 1990 Recommendations of the International Commission on Radiological Protection, ICPR Publ. 60. Annals of ICPR, Vol. 21, No. 1 3, H. Smith, ed. Oxford, U.K.: Pergamon.

Iho, S., T. Takahashi, F. Kura, H. Sugiyama, and T. Hoshino. 1986. Effect of 1,25-dihydroxyvitamin D_3 on in vitro immunoglobin production in human B cells. J. Immunol. 136:4427 4431.

Indraprasit, S., G.V. Alexander, and H.C. Gonick. 1974. Tissue composition of major and trace elements in uremia and hypertension. J. Chronic Dis. 27:135 161.

IOM (Institute of Medicine). 1985. Preventing Low Birthweight. Washington, D.C.: National Academy Press. 284 pp.

Iyengar, G.V. 1989. Elemental Analysis of Biological Systems: Biomedical, Environmental, Compositional, and Methodological Aspects of Trace Elements, Vol. 1. Boca Raton, Fla: CRC Press. 242 pp.

Jacobson, B.E., G. Lockitch, and G. Quigley. 1991. Improved sample preparation for accurate determination of low concentrations of lead in whole blood by graphite furnace analysis. Clin. Chem. 37:515 519.

Jagner, D. 1982. Potentiometric stripping analysis. Analyst 107:593 599.

Jagner, D., M. Josefson, S. Westerlund, and K. Årén. 1981. Simultaneous determination of cadmium and lead in whole blood and in serum by computerized potentiometric stripping analysis. Anal. Chem. 53:1406 1410.

James, A.C. 1978. Lung deposition of sub-micron aerosols calculated as a function of age and breathing rate. Pp. 71 75 in National Radiological Protection Board Annual Research and Development Report. Harwell, U.K.: National Radiological Protection Board, Atomic Energy Research Establishment.

Janghorbani, M. 1984. Stable isotopes in nutrition and food science. Prog. Food Nutr. Sci. 8:303 332.
Janghorbani, M., and B.T.G. Ting. 1989a. Stable isotope tracer applications of ICP-MS. Pp. 115 140 in Applications of Inductively Coupled Plasma Mass Spectrometry, A.R. Date and A.L. Gray, eds. New York: Chapman and Hall.
Janghorbani, M., and B.T.G. Ting. 1989b. Comparison of pneumatic nebulization and hydride generation inductively coupled plasma mass spectrometry for isotopic analysis of selenium. Anal. Chem. 61: 701 708.
Janghorbani, M., B.T.G. Ting, and S.H. Zeisel. 1988. Trace metal research with stable isotope tracers. Curr. Top. Nutr. Dis. 18:545 556.
Johnson, N.E., and K. Tenuta. 1979. Diets and lead blood levels of children who practice pica. Environ. Res. 18:369 376.
Jones, K.W., G. Schidlovsky, F.H. Williams, Jr., R.P. Wedeen, and V. Batuman. 1987. In vivo determination of tibial lead by K-x-ray fluorescence with a Cd-109 source. Pp. 363 373 in In Vivo Body Composition Studies, Proceedings of an International Symposium. K.J. Ellis, S. Yasamura, and W.D. Morgan, eds. Oxford, U.K.: Bocardo Press.
Kalef-Ezra, J.A., D.N. Slatkin, J.F. Rosen, and L. Wielopolski. 1990. Radiation risk to the human conceptus from measurement of maternal tibial bone lead by L-line x-ray fluorescence. Health Phys. 58:217 218.
Kang, H.K., P.F. Infante, and J.S. Carra. 1983. Determination of blood-lead elimination patterns of primary lead smelter workers. J. Toxicol. Environ. Health 11:199 210.
Karalekas, P.C., Jr., C.R. Ryan, and F.B. Taylor. 1983. Control of lead, copper, and iron pipe corrosion in Boston. J. Am. Water Works Assoc. 75:92 95.
Kasprzak, K.S., K.L. Hoover, and L.A. Poirier. 1985. Effects of dietary calcium acetate on lead subacetate carcinogenicity in kidneys of male Sprague-Dawley rats. Carcinogenesis 6:279 282.
Kawada, T., H. Koyama, and T. Suzuki. 1989. Cadmium 'NAG activity' and β_2-microglobulin in the urine of cadmium pigment workers. Br. J. Ind. Med. 46:52 55.
Kawaguchi, H. 1988. Inductively coupled plasma mass spectrometry. Anal. Sci. 4:339 345.

Keating, A.D., J.L. Keating, D.J. Halls, and G.S. Fell. 1987. Determination of lead in teeth by atomic absorption spectrometry with electrothermal atomisation. Analyst 112:1381 1385.

Kehoe, R.A. 1961a. The metabolism of lead in man in health and disease: The normal metabolism of lead. J. R. Inst. Public Health Hyg. 24:81 97.

Kehoe, R.A. 1961b. The metabolism of lead in man in health and disease: The metabolism of lead under abnormal conditions. J. R. Inst. Public Health Hyg. 24:129 143.

Kehoe, R.A. 1961c. The metabolism of lead in man in health and disease: Present hygienic problems relating to the absorption of lead. J. R. Inst. Public Health Hyg. 24:177 203.

Ketterer, M.E., J.J. Reschl, and M.J. Peters. 1989. Multivariate calibration in inductively coupled plasma mass spectrometry. Anal. Chem. 61:2031 2040.

Khera, A.K., D.G. Wibberley, and J.G. Dathan. 1980. Placental and stillbirth tissue lead concentrations in occupationally exposed women. Br. J. Ind. Med. 37:394 396.

Kirkby, H., and F. Gyntelberg. 1985. Blood pressure and other cardiovascular risk factors of long-term exposure to lead. Scand. J. Work Environ. Health 11:15 19.

Klann, E., and K.R. Shelton. 1989. The effect of lead on the metabolism of a nuclear matrix protein which becomes prominent in lead-induced intranuclear inclusion bodies. J. Biol. Chem. 264:16969 16972.

Klein, H., R. Namer, E. Harper, and R. Corbin. 1970. Earthenware containers as a source of fatal lead poisoning. N. Engl. J. Med. 283:669 672.

Kneip, T.J., R.P. Mallon, and N.H. Harley. 1983. Biokinetic modeling for mammalian lead metabolism. NeuroToxicology 4:189 192.

Koh, T.-S., and P.J. Babidge. 1986. A comparison of blood lead levels in dogs from lead-mining, lead-smelting urban and rural island environment. Aust. Vet. J. 63:282 285.

Koo, W.W., P.A. Succop, R.L. Bornschein, S.K. Krug-Wispe, J.J. Steinchen, R.C. Tsang, and O.G. Berger. 1991. Serum vitamin D metabolites and bone mineralization in young children with chronic low to moderate lead exposure. Pediatrics 87:680 687.

Koppenaal, D.W. 1988. Atomic mass spectrometry. Anal. Chem. 60:113R 131R.

Koppenaal, D.W. 1990. Atomic mass spectrometry. Anal. Chem. 62:303R 324R.
Korpela, H., R. Loueniva, E. Yrjanheikki, and A. Kauppila. 1986. Lead and cadmium concentrations in maternal and umbilical cord blood, amniotic fluid, placenta, and amniotic membranes. Am. J. Obstet. Gynecol. 155:1086 1089.
Kostial, K., D. Kello, S. Jugo, I. Rabar, and T. Maljkovic. 1978. Influence of age on metal metabolism and toxicity. Environ. Health Perspect. 25:81 86.
Kromhout, D. 1988. Blood lead and coronary heart disease risk among elderly men in Zutphen, the Netherlands. Environ. Health Perspect. 78:43 46.
Kromhout, D., A.A.E. Wibowo, R.F.M. Herber, L.M. Dalderup, H. Heerdink, C. de Lezenne Coulander, and R.L. Zielhuis. 1985. Trace metals and coronary heart disease risk indicators in 152 elderly men (the Zutphen study). Am. J. Epidemiol. 122:378 385.
Lal, B., R.C. Murthy, M. Anand, S.V. Chandra, R. Kumar, O. Tripathi, and R.C. Srimal. 1991. Cardiotoxicity and hypertension in rats after oral lead exposure. Drug Chem. Toxicol. 14:305 318.
Lamola, A.A., M. Joselow, and T. Yamane. 1975. Zinc protoporphyrin (ZPP): A simple, sensitive fluorometric screening test for lead poisoning. Clin. Chem. 21:93 97.
Lancranjan, I., H.I. Popescu, O. Gavanescu, I. Klepsch, and M. Serbanescu. 1975. Reproductive ability of workmen occupationally exposed to lead. Arch. Environ. Health 30:396 401.
Landis, J.R., and K.M. Flegal. 1988. A generalized Mantel-Haenszel analysis of the regression of blood pressure on blood lead using NHANES II data. Environ. Health Perspect. 78:35 41.
Landrigan, P.J. 1989. Toxicity of lead at low dose. Br. J. Ind. Med. 46:593 596.
Landrigan, P.J., E.L. Baker, Jr., R.G. Feldman, D.H. Cox, K.V. Eden, W.A. Orenstein, J.A. Mather, A.J. Yankel, and I.H. von Lindern. 1976. Increased lead absorption with anemia and slowed nerve conduction in children near a lead smelter. J. Pediatr. 89: 904 910.
Landrigan, P.J., E.L. Baker, Jr., J.S. Himmelstein, G.F. Stein, J.P. Weddig, and W.E. Straub. 1982. Exposure to lead from the Mystic River Bridge: The dilemma of deleading. N. Engl. J. Med. 306: 673 676.

Lane, R.E. 1949. The care of the lead worker. Br. J. Ind. Med. 6:125 143.
Lansdown, R., W. Yule, M.-A. Urbanowicz, and J. Hunter. 1986. The relationship between blood-lead concentrations, intelligence, attainment and behaviour in a school population: The second London study. Int. Arch. Occup. Environ. Health 57:225 235.
Lauwers, M.-C., R.C. Hauspie, C. Susanne, and J. Verheyden. 1986. Comparison of biometric data of children with high and low levels of lead in the blood. Am. J. Phys. Anthropol. 69:107 116.
Lee, J.A., and J.H. Tallis. 1973. Regional and historical aspects of lead pollution in Britain. Nature 245:216 218.
Legge, T.M. 1901. Industrial lead poisoning. J. Hyg. 1:96 108.
Levin, R. 1986. Reducing Lead in Drinking Water: A Benefit Analysis. EPA-230/09-86-019. Washington, D.C.: U.S. Environmental Protection Agency.
Levin, R. 1987. Lead in U.S. public drinking water: The benefits of reducing that exposure. Pp. 215 217 in International Conference on Heavy Metals in the Environment, Vol. 2, S.E. Lindberg and T.C. Hutchinson, eds. Edinburgh: CEP Consultants.
Levin, R., and M.R. Schock. 1991. The use of pipe loop tests for corrosion control diagnostics. Pp. 697 723 in Proceedings of the Water Quality Technology Conference: Advances in Water Analysis and Treatment, San Diego, Nov. 11 15, 1990. Denver, Colo.: American Water Works Association.
Leviton, A., D. Bellinger, E. Allred, M. Rabinowitz, H. Needleman, and S. Schoenbaum. 1993. Pre- and postnatal low-level lead exposure and children's dysfunction in school. Environ. Res. 60:30 43.
Lieberherr, M. 1987. Effects of vitamin D_3 metabolites on cytosolic free calcium in confluent mouse osteoblasts. J. Biol. Chem. 262: 13168 13173.
Lilis, R., J. Eisinger, W. Blumberg, A. Fischbein, and I.J. Selikoff. 1978. Hemoglobin, serum iron, and zinc protoporphyrin in lead-exposed workers. Environ. Health Perspect. 25:97 102.
Lilley, S.G., T.M. Florence, and J.L. Stauber. 1988. The use of sweat to monitor lead absorption through the skin. Sci. Total Environ. 76:267 278.
Lindberg, S.E., and R.C. Harriss. 1981. The role of atmospheric deposition in an eastern U.S.A. deciduous forest. Water Air Soil Pollut. 16:13 31.

Lindh, U., D. Brune, and G. Nordberg. 1978. Micorprobe analysis of lead in human femur by proton induced x-ray emission (PIXE). Sci. Total Environ. 10:31 37.

Lin-Fu, J.S. 1980. Lead poisoning and undue lead exposure in children: History and current status. Pp. 5 16 in Low Level Lead Exposure: The Clinical Implications of Current Research, H.L. Needleman, ed. New York: Raven.

Lin-Fu, J.S. 1982. The evolution of childhood lead poisoning as a public health problem. Pp. 1 10 in Lead Absorption in Children: Management, Clinical, and Environmental Aspects, J.J. Chisolm, Jr., and D.M. O'Hara, eds. Baltimore: Urban & Schwarzenberg.

Linton, R.W., D.F.S. Natusch, R.L. Solomon, and C.A. Evans, Jr. 1980. Physicochemical characterization of lead in urban dusts: A microanalytical approach to lead tracing. Environ. Sci. Technol. 14:159 164.

Linton, R.W., S.R. Bryan, P.F. Schmidt, and D.P. Griffis. 1985. Comparison of laser and ion microprobe detection sensitivity for lead in biological microanalysis. Anal. Chem. 57:440 443.

Livett, E.A., J.A. Lee, and J.H. Tallis. 1979. Lead, zinc and copper analysis of British blanket peats. J. Ecol. 67:865 891.

Lodding, A. 1988. Secondary ion mass spectrometry. Pp. 125 171 in Inorganic Mass Spectrometry, F. Adams, R. Gijbels, and R. Van Grieken, eds. New York: John Wiley & Sons.

Lolin, Y., and P. O'Gorman. 1986. δ-Aminolaevulinic acid dehydratase as an index of the presence and severity of lead poisoning in acute and chronic lead exposure. Ann. Clin. Biochem. 23:521 528.

Lolin, Y., and P. O'Gorman. 1988. An intra-erythrocytic low molecular weight lead-binding protein in acute and chronic lead exposure and its possible protective role in lead toxicity. Ann. Clin. Biochem. 25:688 697.

Long, G.I., J.F. Rosen, and J.G. Pounds. 1990. Cellular lead toxicity and metabolism in primary and clonal osteoblastic bone cells. Toxicol. Appl. Pharmacol. 102:346 361.

Longerich, H.P., B.J. Fryer, and D.F. Strong. 1987. Determination of lead isotope ratios by inductively coupled plasma-mass spectrometry (ICP-MS). Spectrochim. Acta 42B:39 48.

Lorimer, G. 1886. Saturnine gout and its distinguishing marks. Br. Med. J. 2(1334):163.

Low, J.A., and R.S. Galbraith. 1974. Pregnancy characteristics of intrauterine growth retardation. Obstet. Gynecol. 44:122 126.

Lozoff, B., A.W. Wolf, J.J. Urrutia, and F.E. Viteri. 1985. Abnormal behavior and low developmental test scores in iron-deficient anemic infants. J. Dev. Behav. Pediatr. 6:69 75.

Lozoff, B., G.M. Brittenham, A.W. Wolf, D.K. McClish, P.M. Kuhnert, E. Jimenez, R. Jimenez, L.A. Mora, I. Gomez, and D. Krauskoph. 1987. Iron deficiency anemia and iron therapy effects on infant development test performance. Pediatrics 79:981 985; Erratum, 1988, 81:683.

Lyngbye, T., O. Hansen, A. Trillingsgaard, I. Beese, and P. Grandjean. 1990. Learning disabilities in children: Significance of low-level lead-exposure and confounding factors. Acta Paediatr. Scand. 79:352 360.

Lyons, T.D.B., G.S. Fell, R.C. Hutton, and A.N. Eaton. 1988. Elimination of chloride interference on the determination of selenium in serum by inductively coupled plasma mass spectrometry. J. Anal. At. Spectrom. 3:601 603.

Mackie, A.C., R. Stephens, A. Townsend, and H.A. Waldron. 1977. Tooth lead levels in Birmingham children. Arch. Environ. Health 32:178 185.

Mahaffey, K.R., ed. 1985. Dietary and Environmental Lead: Human Health Effects. Amsterdam: Elsevier. 459 pp.

Mahaffey, K.R. 1992. Exposure to lead in childhood: The importance of prevention [editorial]. N. Engl. J. Med. 327:1308 1309.

Mahaffey, K.R., and J.L. Annest. 1986. Association of erythrocyte protoporphyrin with blood lead level and iron status in the Second National Health and Nutrition Examination Survey, 1976 1980. Environ. Res. 41:327 338.

Mahaffey, K.R., and B.A. Fowler. 1977. Effects of concurrent administration of lead, cadmium, and arsenic in the rat. Environ. Health Perspect. 19:165 171.

Mahaffey, K.R., and I.A. Michaelson. 1980. The interaction between lead and nutrition. Pp. 159 200 in Low Level Lead Exposure: The Clinical Implications of Current Research, H.L. Needleman, ed. New York: Raven.

Mahaffey, K.R., J.D.. Haseman, R.A. Goyer. 1973. Dose-response to lead ingestion in rats on low dietary calcium. J. Lab. Clin. Med. 83:92 100.

Mahaffey, K.R., J.L. Annest, H.E. Barbano, and R.S. Murphy. 1979. Preliminary analysis of blood lead concentrations for children and adults: HANES II, 1976 1978. Trace Subst. Environ. Health 13:37 51.
Mahaffey, K.R., S.G. Capar, B.C. Gladen, and B.A. Fowler. 1981. Concurrent exposure to lead, cadmium, and arsenic: Effects on toxicity and tissue metal concentrations in the rat. J. Lab. Clin. Med. 98:463 481.
Mahaffey, K.R., J.L. Annest, J. Roberts, and R.S. Murphy. 1982a. National estimates of blood lead levels: United States 1976 1980. New Engl. J. Med. 307:573 579.
Mahaffey, K.R., J.F. Rosen, R.W. Chesney, J.T. Peeler, C.M. Smith, and H.F. DeLuca. 1982b. Association between age, blood lead concentration, and serum 1,25-dihydroxycholecalciferol levels in children. Am. J. Clin. Nutr. 35:1327 1331.
Mahaffey-Six, K.R., and R.A. Goyer. 1972. The influence of iron deficiency on tissue content and toxicity of ingested lead in the rat. J. Lab. Clin. Med. 79:128 136.
Mahomed, F.A. 1891. Chronic Bright's disease without albuminuria. Guy's Hosp. Rep. 25(Ser. 3):295 416.
Maines, M.D., and A. Kappas. 1977. Enzymes of heme metabolism in the kidney: Regulation by trace metals which do not form heme complexes. J. Exp. Med. 146:1286 1293.
Major, R.H. 1945. Classic Descriptions of Disease, 3rd Ed. Springfield, Ill.: Charles C Thomas.
Manalis, R.S., and G.P. Cooper. 1973. Presynaptic and postsynaptic effects of lead at the frog neuromuscular junction. Nature 243:354 355.
Manton, W.I. 1977. Sources of lead in blood: Identification by stable isotopes. Arch. Environ. Health 32:149 159.
Manton, W.I. 1985. Total contribution of airborne lead to blood lead. Br. J. Ind. Med. 42:168 172.
Manton, W.I., and J.D. Cook. 1984. High accuracy (stable isotope dilution) measurements of lead in serum and cerebrospinal fluid. Br. J. Ind. Med. 41:313 319.
Manton, W.I., and C.R. Malloy. 1983. Distribution of lead in body fluids after ingestion of soft solder. Br. J. Ind. Med. 40:51 57.
Marcus, A.H. 1985. Testing alternative nonlinear kinetic models in compartmental analysis. Pp. 259 267 in Mathematics and Comput-

ers in Biomedical Applications, J. Eisenfeld and C. DeLisi, eds. Amsterdam: Elsevier/North-Holland.

Marcus, A.H., and J. Schwartz. 1987. Dose-response curves for erythrocyte protoporphyrin vs. blood lead: Effects of iron status. Environ. Res. 44:221 227.

Marden, P.M., D.W. Smith, and M.J. McDonald. 1964. Congenital anomalies in the newborn infant, including minor variations. J. Pediatr. 64:357 371.

Marecek, J., I. Shapiro, A. Burke, S. Katz, and M. Hediger. 1983. Low-level lead exposure in childhood influences neuropsychological performance. Arch. Environ. Health 38:355 359.

Markovac, J., and G.W. Goldstein. 1988a. Lead activates protein kinase C in immature rat brain microvessels. Toxicol. Appl. Pharmacol. 96:14 23.

Markovac, J., and G.W. Goldstein. 1988b. Picomolar concentrations of lead stimulate brain protein kinase C. Nature 334:71 73.

Markowitz, M.E., and J.F. Rosen. 1981. Zinc (Zn) and copper (Cu) metabolism in $CaNa_2EDTA$-treated children with plumbism. Pediatr. Res. 15:635.

Markowitz, M.E., and J.F. Rosen. 1984. Assessment of body lead stores in children: Validation of an 8-hour $CaNa_2EDTA$ provocative test. J. Pediatr. 104:337 342.

Markowitz, M.E., and J.F. Rosen. 1990. Higher bone lead content in black compared to nonblack children [abstract]. Pediatr. Res. 27: 95A.

Markowitz, M.E., and H.L. Weinberger. 1990. Immobilization-related lead toxicity in previously lead-poisoned children. Pediatrics 86:455 457.

Marshall, J. 1988. The ICP Is it the real thing? Anal. Proc. (London) 25:238 240.

Matson, W.R., R.M. Griffin, and G.B. Schreiber. 1970. Rapid sub-nanogram simultaneous analysis of Zn, Cd, Pb, Cu, Bi, and Ti. Trace Subst. Environ. Health 4:396 406.

Mayer-Popken, O., W. Denkhaus, and H. Konietzko. 1986. Lead content of fetal tissues after maternal intoxication. Arch. Toxicol. 58:203 204.

McBride, W.G., B.P. Black, and B.J. English. 1982. Blood lead levels and behaviour of 400 pre-school children. Med. J. Aust. 2:26 29.

McElvaine, M.D., H.G. Orbach, S. Binder, L.A. Blanksma, E.F. Maes, and R.M. Krieg. 1991. Evaluation of the erythrocyte protoporphyrin test as a screen for elevated blood lead levels. J. Pediatr. 119:548 550.
McKhann, C.F. 1926. Lead poisoning in children. Am. J. Dis. Child. 2:386 392.
McMichael, A.J., G.V. Vimpani, E.F. Robertson, P.A. Baghurst, and P.D. Clark. 1986. The Port Pirie cohort study: Maternal blood lead and pregnancy outcome. J. Epidemiol. Commun. Health 40: 18 25.
McMichael, A., P. Baghurst, N. Wigg, G. Vimpani, E. Robertson, and R. Roberts. 1988. Port Pirie cohort study: Environmental exposure to lead and children's abilities at the age of four years. N. Engl. J. Med. 319:468 475.
McMichael, A., P. Baghurst, G. Vimpani, E. Robertson, N. Wigg, and S.-L. Tong. 1992. Sociodemographic factors modifying the effect of environmental lead on neuropsychological development in early childhood. Neurotoxicol. Teratol. 14:321 327.
Mehani, S. 1966. Lead retention by the lungs of lead-exposed workers. Ann. Occup. Hyg. 9:165 171.
Meredith, P.A., M.R. Moore, B.C. Campbell, G.G. Thompson, and A. Goldberg. 1978. Delta-aminolaevulinic acid metabolism in normal and lead-exposed humans. Toxicology 9:1 9.
Merril, C.R., M.L. Dunau, and D. Goldman. 1981. A rapid sensitive silver stain for polypeptides in polyacrylamide gels. Anal. Biochem. 110:201 207.
Meyer, B.R., A. Fischbein, K. Rosenman, Y. Lerman, D.E. Drager, and M.M. Reidenberg. 1984. Increased urinary enzyme excretion in workers exposed to nephrotoxic chemicals. Am. J. Med. 76:989 998.
Mezzetti, G., M.G. Monti, L.P. Casolo, G. Piccinini, and M.S. Moruzzi. 1988. 1,25-Dihydroxycholecalciferol-dependent calcium uptake by mouse mammary gland in culture. Endocrinology 122:389 394.
Milar, C.R., and P. Mushak. 1982. Lead-contaminated house dust: Hazard, measurement, and decontamination. Pp. 143 152 in Lead Absorption in Children: Management, Clinical, and Environmental Aspects, J.J. Chisolm, Jr., and D.M. O'Hara, eds. Baltimore, Md.: Urban & Schwarzenberg.
Milar, C.R., S.R. Schroeder, P. Mushak, J.L. Dolcourt, and L.D.

Grant. 1980. Contributions of the care-giving environment to increased lead burden of children. Am. J. Mental Deficiency 84: 339 344.

Miller, D.T., D.C. Paschal, E.W. Gunther, P.E. Stroud, and J.D'Angelo. 1987. Determination of lead in blood using electrothermal atomisation atomic absorption spectrometry with a L'vov platform and matrix modifier. Analyst 112:1701 1704.

Minnema, D.J. 1989. Neurochemical alterations in lead intoxication: An overview. Comments Toxicol. 3:207 224.

Mistry, P., G.W. Lucier, and B.A. Fowler. 1985. High-affinity lead binding proteins in rat kidney cytosol mediate cell-free nuclear translocation of lead. J. Pharmacol. Exp. Ther. 232:462 469.

Mistry, P., C. Mastri, and B.A. Fowler. 1986. Influence of metal ions on renal cytosolic lead-binding proteins and nuclear uptake of lead in the kidney. Biochem. Pharmacol. 35:711 713.

Mistry, P., C. Mastri, and B.A. Fowler. 1987. Lead-induced alterations of renal gene expression within subcellular compartment [abstract]. Toxicologist 7:78.

Möller, L., and T.S. Kristensen. 1992. Blood lead as a cardiovascular risk factor. Am. J. Epidemiol. 136:1091 1100.

Montasser, A., S. Chan, and D.W. Koppenaal. 1987. Inductively coupled helium plasma as an ion source for mass spectrometry. Anal. Chem. 59:1240 1242.

Moore, J.F., R.A. Goyer, and M. Wilson. 1973. Lead-induced inclusion bodies: Solubility, amino-acid content and relationship to residual acidic nuclear proteins. Lab. Invest. 29:488 494.

Moore, M.R. 1985. Uptake of lead from water. Adv. Mod. Environ. Toxicol. 9:259 270.

Moore, M.R. 1988. Haematological effects of lead. Sci. Total Environ. 71:419 431.

Moore, M.R., M.A. Hughes, and D.J. Goldberg. 1979. Lead absorption in man from dietary sources: The effect of cooking upon lead concentrations of certain foods and vegetables. Int. Arch. Occup. Environ. Health. 44:81 90.

Moore, M.R., P.A. Meredith, W.S. Watson, D.J. Sumner, M.K. Taylor, and A. Goldberg. 1980. The percutaneous absorption of lead-203 in humans from cosmetic preparations containing lead acetate, as assessed by whole-body counting and other techniques. Food Cosmet. Toxicol. 18:399 405.

Moore, M.R., A. Goldberg, W.M. Fyfe, R.A. Low, and W.N. Richards. 1981. Lead in water in Glasgow: A story of success. Scot. Med. J. 26:354 355.
Moore, M.R., A. Goldberg, S.J. Pocock, P.A. Meredith, I.M. Stewart, H. McAnespie, R. Lees, and A. Low. 1982. Some studies of maternal and infant lead exposure in Glasgow. Scott. Med. J. 27:113 122.
Moore, M.R., W.N. Richards, and J.G. Sherlock. 1985. Successful abatement of lead exposure from water supplies in the west of Scotland. Environ. Res. 38:67 76.
Moore, M.R., A. Goldberg, and A.A. Yeung-Laiwah. 1987. Lead effects on the heme biosynthetic pathway. Relationship to toxicity. Ann. N.Y. Acad. Sci. 514:191 203.
Mooty, J., C.E. Ferand, Jr., and P. Harris. 1975. Relationship of diet to lead poisoning in children. Pediatrics 55:636 639.
Moreau, T., G. Orssaud, B. Juguet, and G. Busquet. 1982. Blood lead levels and arterial pressure: Initial results of a cross sectional study of 431 male subjects [letter in French]. Rev. Epidemiol. Sante Publique 30:395 397.
Morgan, W.D., S.J. Ryde, S.J. Jones, R.M. Wyatt, I.R. Hainsworth, S.S. Cobbold, C.J. Evans, and R.A. Braithwaite. 1990. In vivo measurements of cadmium and lead in occupationally-exposed workers and an urban population. Biol. Trace Elem. Res. 26-27:407 414.
Morrow, P.E., H. Beiter, F. Amato, and F.R. Gibb. 1980. Pulmonary retention of lead: An experimental study in man. Environ. Res. 21:373 384.
Morrow-Tlucak, M., and C.B. Ernhart. 1987. The relationship of low level lead exposure and language development in the pre-school years. Pp. 57 59 in International Conference on Heavy Metals in the Environment, Vol. 1, S.E. Lindberg and T.C. Hutchinson, eds. Edinburgh: CEP Consultants.
Mueller, P.W., S.J. Smith, K.K. Steinberg, and M.J. Thun. 1989. Chronic renal tubular effects in relation to urine cadmium levels. Nephron 52:45 54.
Munro, S., L. Ebdon, and D.J. McWeeny. 1986. Application of inductively coupled plasma mass spectrometry (ICP-MS) for trace metal determination in foods. J. Anal. At. Spectrom. 1:211 219.
Murozumi, M., T.J. Chow, and C. Patterson. 1969. Chemical concentrations of pollutant lead aerosols, terrestial dusts, and sea salts in

Greenland and Antarctic snow strata. Geochim. Cosmochim. Acta 33:1247 1294.

Murphy, M.J., J.H. Graziano, D. Popovac, J.K. Kline, A. Mehmeti, P. Factor-Litvak, G. Ahmedi, P. Shrout, B. Rajovic, D.U. Nenezic, and Z.A. Stein. 1990. Past pregnancy outcomes among women living in the vicinity of a lead smelter in Kosovo, Yugoslavia. Am. J. Public Health 80:33 35.

Mushak, P. 1989. Biological monitoring of lead exposure in children: Overview of selected biokinetic and toxicological issues. Pp. 129 145 in Lead Exposure and Child Development: An International Assessment, M.A. Smith, L.D. Grant, and A.I. Sors, eds. Dordrecht, The Netherlands: Kluwer Academic.

Mushak, P. 1992. The monitoring of human lead exposure. Pp. 45 64 in Human Lead Exposure, H.L. Needleman, ed. Boca Raton, Fla.: CRC Press.

Mushak, P. In press. New directions in the toxicokinetics of human lead exposure. NeuroToxicology.

Mushak, P., and A.F. Crocetti. 1989. Determination of numbers of lead-exposed American children as a function of lead source: Integrated summary of a report to the U.S.Congress on childhood lead poisoning. Environ. Res. 50:210 229.

Naidoo, D.V., and J. Moodley. 1980. A survey of hypertension in pregnancy at the King Edward VIII Hospital, Durban, South Africa. S. Afr. Med. J. 58:556 559.

Nakhoul, F., L.H. Kayne, N. Brautbar, M.S. Hu, A. McDonough, and P. Eggena. 1992. Rapid hypertensinogenic effect of lead: Studies in the spontaneously hypertensive rat. Toxicol. Ind. Health 8:89 102.

NCEMCH (National Center for Education in Maternal and Child Health). 1989. Childhood Lead Poisoning Prevention: A Resource Directory. Washington, D.C.: National Center for Education in Maternal and Child Health.

NCRP (National Council on Radiation Protection and Measurements). 1989. Concepts, units, and quantities, p. 6, and Appendix A, pp. 73 74 in Exposure of the U.S. Population from Diagnostic Medical Radiation, Report 100. Bethesda, Md.: National Council on Radiation Protection and Measurements.

Needleman, H.L., and C.A. Gatsonis. 1990. Low level lead exposure and the IQ of children: A meta-analysis of modern studies. J. Am. Med. Assoc. 263:673 678.

Needleman, H.L., and I.M. Shapiro. 1974. Dentine lead levels in asymptomatic Philadelphia school children: Subclinical exposure in high and low risk groups. Environ. Health Perspect. 7:27 31.

Needleman, H.L., C. Gunnoe, A. Leviton, R. Reed, H. Peresie, C. Maher, and P. Barrett. 1979. Deficits in psychologic and classroom performance of children with elevated dentine lead levels. N. Engl. J. Med. 300:689 695.

Needleman, H.L., A. Leviton, and D. Bellinger. 1982. Lead-associated intellectual deficits [letter]. N. Engl. J. Med. 306:367.

Needleman, H.L., M. Rabinowitz, A. Leviton, S. Linn, and S. Schoenbaum. 1984. The relationship between prenatal exposure to lead and congenital anomalies. J. Am. Med. Assoc. 251:2956 2959.

Needleman, H.L., A. Schell, D. Bellinger, A. Leviton, and E.N. Allred. 1990. The long-term effects of childhood exposure to low doses of lead: An 11-year follow-up report. N. Engl. J. Med. 322:83 88.

Nemere, I., and A.W. Norman. 1988. 1,25-Dihydroxyvitamin D_3-mediated vesicular transport of calcium in intestine: Time-course studies. Endocrinology 122:2962 2969.

Neri, L.C., D. Hewitt, and B. Orser. 1988. Blood lead and blood pressure: Analysis of cross-sectional and longitudinal data from Canada. Environ. Health Perspect. 78:123 126.

Ng, A., and C.C. Patterson. 1981. Natural concentrations of lead in ancient Arctic and Antarctic ice. Geochim. Cosmochim. Acta 45: 2109 2121.

Ng, A., and C.C. Patterson. 1982. Changes of lead and barium with time in California (USA) off-shore basin sediments. Geochim. Cosmochim. Acta 46:2307 2322.

Nieboer, E., and A.A. Jusys. 1983. Contamination control in routine ultratrace analysis of toxic metals. Pp. 3 16 in Chemical Toxicology and Clinical Chemistry of Metals, S.S. Brown and J. Savory, eds. New York: Academic.

Nielsen, T., K.A. Jensen, and P. Grandjean. 1978. Organic lead in normal human brains. Nature 274:602 603.

Niklowitz, W.J., and T.I. Mandybur. 1975. Neurofibrillary changes

following childhood lead encephalopathy: Case report. J. Neuropathol. Exp. Neurol. 34:445 455.
Nilsson, U., R. Attewell, J.-O. Christoffersson, A. Schütz, L. Ahlgren, S. Skerfving, S. Mattsson. 1991. Kinetics of lead in bone and blood after end of occupational exposure. Pharmacol. Toxicol. 69:477 484.
Nordstrom, S., L. Beckman, and I. Nordenson. 1978a. Occupational and environmental risks in and around a smelter in northern Sweden: Part 1. Variations in birth weight. Hereditas 88:43 46.
Nordstrom, S., L. Beckman, and I. Nordenson. 1978b. Occupational and environmental risks in and around a smelter in northern Sweden. Part 3. Frequencies of spontaneous abortion. Hereditas 88:51 54.
Norman, A.W. 1990. Intestinal calcium absorption: A vitamin D-hormone-mediated adaptive response. Am. J. Clin. Nutr. 51:290 300.
Nozaki, K. 1966. Method for studies on inhaled particles in the human respiratory system and retention of lead fumes. Ind. Health 4:118 128.
NRC (National Research Council). 1972. Lead: Airborne Lead in Perspective. Washington, D.C.: National Academy Press. 330 pp.
NRC (National Research Council). 1976. Recommendations for the Prevention of Lead Poisoning in Children. Washington, D.C.: National Academy Press. 65 pp.
NRC (National Research Council). 1980. Lead in the Human Environment. Washington, D.C.: National Academy Press. 525 pp.
NRC (National Research Council). 1986. Drinking Water and Health. Vol. 6. Washington, D.C.: National Academy Press. 457 pp.
NRC (National Research Council). 1989a. Biological Markers in Pulmonary Toxicology. Washington, D.C.: National Academy Press. 179 pp.
NRC (National Research Council). 1989b. Biological Markers in Reproductive Toxicology. Washington, D.C.: National Academy Press. 395 pp.
NRC (National Research Council). 1990. Health Effects of Exposure to Low Levels of Ionizing Radiation: BEIR V. Washington, D.C.: National Academy Press. 421 pp.
Nriagu, J.O. 1978. Lead in the atmosphere. Pp. 137 184 in The Biogeochemistry of Lead in the Environment. Part A. Ecological Cycles. Topics in Environmental Health, Vol. 1A, J.O. Nriagu, ed. Amsterdam: Elsevier/North-Holland.

Nriagu, J.O. 1983a. Lead exposure and lead poisoning. Pp. 309 424 in Lead and Lead Poisoning in Antiquity. New York: John Wiley & Sons.

Nriagu, J.O. 1983b. Saturnine gout among Roman aristocrats. Did lead poisoning contribute to the fall of the empire? N. Engl. J. Med. 308:660 663.

Nriagu, J.O. 1985a. Cupellation: The oldest quantitative chemical process. J. Chem. Ed. 62:668 672.

Nriagu, J.O. 1985b. Historical perspective on the contamination of food and beverages with lead. Pp. 1 41 in Dietary and Environmental Lead: Human Health Effects, K.R. Mahaffey, ed. Amsterdam: Elsevier.

Nriagu, J.O. 1989. A global assessment of natural sources of atmospheric trace metals. Nature 338:47 49.

Nriagu, J.O. 1990. The rise and fall of leaded gasoline. Sci. Total Environ. 92:13 28.

Nriagu, J.O., and J.M. Pacyna. 1988. Quantitative assessment of worldwide contamination of air, water and soils by trace metals. Nature 333:134 139.

Nürnberg, H.W. 1983. Potentialities and applications of voltammetry in the analysis of toxic trace metals in body fluids. Pp. 209 232 in Analytical Techniques for Heavy Metals in Biological Fluids, S. Facchetti, ed. Amsterdam: Elsevier.

O'Farrell, P.H. 1975. High resolution two-dimensional poly acrylamide gel electrophoresis of proteins. J. Biol. Chem. 250:4007 4021.

O'Flaherty, E.J., P.B. Hammond, and S.I. Lerner. 1982. Dependence of apparent blood lead half-life on the length of previous lead exposure in humans. Fundam. Appl. Toxicol. 2:49 54.

Ohanian, E.V. 1986. Health effects of corrosion products in drinking water. Trace Subst. Environ. Health 20:122 138.

Okayama, A., Y. Ogawa, K. Miyajima, M. Hirata, T. Yoshida, T. Tabuchi, K. Sugimoto, and K. Morimoto. 1989. A new HPLC fluorimetric method to monitor urinary delta-aminolevulinic acid (ALA-U) levels in workers exposed to lead. Int. Arch. Occup. Environ. Health 61:297 302; Erratum 61:426.

Okazaki, H., S.M. Aronson, D.J. DiMaio, and J.E. Olvera. 1963. Acute lead encephalopathy of childhood: Histologic and chemical studies, with particular reference to angiopathic aspects. Trans. Am. Neurol. Assoc. 88:248 250.

Oliver, T. 1911. A lecture on lead poisoning and the race. Br. Med. J. 1:1096 1098.
Olson, K.W., and R.K. Skogerboe. 1975. Identification of soil lead compounds from automotive sources. Environ. Sci. Technol. 9:227 230.
Orssaud, G., J.R. Claude, T. Moreau, J. Lellouch, B. Juget, and B. Festy. 1985. Blood lead concentration and blood pressure [letter]. Br. Med. J. 290:244.
Oskarsson, A., and B.A. Fowler. 1985a. Effects of lead inclusion bodies on subcellular distribution of lead in rat kidney: The relationship to mitochondrial function. Exp. Mol. Pathol. 43:397 408.
Oskarsson, A., and B.A. Fowler. 1985b. Effects of lead on the heme biosynthetic pathway in rat kidney. Exp. Mol. Pathol. 43:409 417.
Oskarsson, A., and B.A. Fowler. 1987. Alterations in renal heme biosynthesis during metal nephrotoxicity. Ann. N.Y. Acad. Sci. 514:268 277.
Oskarsson, A., K.S. Squibb, and B.A. Fowler. 1982. Intracellular binding of lead in the kidney: The partial isolation and characterization of postmitochondrial lead-binding components. Biochem. Biophys. Res. Commun. 104:290 298.
Osterloh, J., and C.E. Becker. 1986. Pharmacokinetics of $CaNa_2$-EDTA and chelation of lead in renal failure. Clin. Pharmacol. Ther. 40:686 693.
Osteryoung, J. 1988. Electrothermal methods of analyses. Pp. 243 267 in Methods in Enzymology, Metallobiochemistry, Part A, J.F. Riordan and B.L. Vallee, eds. New York: Academic.
Ott, M.G., M.J. Teta, and H.L. Greenberg. 1989. Assessment of exposure to chemicals in a complex work environment. Am. J. Ind. Med. 16:617 630.
Ottaway, J.M. 1983. Heavy metals determinations by atomic absorption and emission spectrometry. Pp. 171 208 in Analytical Techniques for Heavy Metals in Biological Fluids, S. Facchetti, ed. Amsterdam: Elsevier.
Otto, D.A. 1989. Electrophysiological assessment of sensory and cognitive function in children exposed to lead: A review. Pp. 279 292 in Lead Exposure and Child Development: An International Assessment, M.A. Smith, L.D. Grant, and A.I. Sors, eds. Dordrecht, The Netherlands: Kluwer Academic.

Page, A.L., T.J. Ganje, and M.S. Joshi. 1971. Lead quantities in plants, soil, and air near some major highways in southern California. Hilgardia 41:1 31.

Page, R.A., P.A. Cawse, and S.J. Baker. 1988. The effect of reducing petrol lead on airborne lead in Wales, U.K. Sci. Total Environ. 68:71 77.

Paglia, D.E., and W.N. Valentine. 1975. Characteristics of a pyrimidine-specific 5' nucleotidase in human erythrocytes. J. Biol. Chem. 250:7973 7979.

Palca, J. 1991. News & Comment. Get-the-lead-out guru challenged. Science 253:842 844.

Palmer, K.T., and C.L. Kucera. 1980. Lead contamination of sycamore and soil from lead mining and smelting operations in eastern Missouri. J. Environ. Qual. 9:106 111.

Panova, Z. 1972. Early changes in the ovarian function of women in occupational contact with inorganic lead. Pp. 161 166 in Works of the United Research Institute of Hygiene and Industrial Safety. Sofia, Bulgaria: United Research Institute of Hygiene and Industrial Safety.

Parke, D.V. 1987. The role of enzymes in protection mechanisms for human health. Regul. Toxicol. Pharmacol. 7:222 235.

Paterson, L.J., G.M. Raab, R. Hunter, D. Laxen, M. Fulton, G.S. Fell, D.J. Halls, and P. Sutcliffe. 1988. Factors influencing lead concentrations in shed deciduous teeth. Sci. Total Environ. 74:219 233.

Patterson, C.C. 1980. An alternative perspective Lead pollution in the human environment: Origin, extent, and significance. Pp. 265 349 in Lead in the Human Environment. Washington, D.C.: National Academy Press.

Patterson, C.C., and D.M. Settle. 1976. The reduction of orders of magnitude errors in lead analyses of biological materials and natural waters by evaluating and controlling the extent and sources of industrial lead contamination introduced during sample collecting, handling, and analysis. Pp. 321 351 in Accuracy in Trace Analysis: Sampling, Sample Handling, and Analysis, Vol. 1, P. LaFleur, ed. National Bureau of Standards Special Publ. 422. Washington, D.C.: U.S. Department of Commerce.

Patterson, C.C., and D.M. Settle. 1987. Review of data on aeolian

fluxes of industrial and natural lead to the lands and seas in remote regions on a global scale. Mar. Chem. 22:137 162.
Patterson, C.C., H. Shirahata, and J.E. Ericson. 1987. Lead in ancient human bones and its relevance to historical developments of social problems with lead. Sci. Total Environ. 61:167 200.
Patterson, C., J. Ericson, M. Manea-Krichten, and H. Shirahata. 1991. Natural skeletal levels of lead in *Homo sapiens sapiens* uncontaminated by technological lead. Sci. Total Environ. 107:205 236.
Paul, C. 1860. Study of the effect of slow lead intoxication on the product of conception [in French]. Arch. Gen. Med. 15:513 533.
Pearson, D.T., and K.N. Dietrich. 1985. The behavioral toxicology and teratology of childhood: Models, methods, and implications for intervention. NeuroToxicology 6:165 182.
Pentschew, A. 1965. Morphology and morphogenesis of lead encephalopathy. Acta Neuropathol. 5:133 160.
Perlstein, M.A., and R. Attala. 1966. Neurologic sequalae of plumbism in children. Clin. Pediatr. 5:292 298.
Perry, H.M., Jr., and M.W. Erlanger. 1978. Pressor effects of chronically feeding cadmium and lead together. Trace Subst. Environ. Health 12:268 275.
Phalen, R.F., M.J. Oldham, C.B. Beaucage, T.T. Crocker, and J.D. Mortensen. 1985. Postnatal enlargement of human tracheobronchial airways and implications for particle deposition. Anat. Rec. 212: 368 380.
Piccinini, F., L. Favalli, and M.C. Chiari. 1977. Experimental investigations on the contraction induced by lead in arterial smooth muscle. Toxicology 8:43 51.
Pickford, C.J., and R.M. Brown. 1986. Comparison of ICP-MS with ICP-ES: Detection power and interference effects experienced with complex matrixes. Spectrochim. Acta 41B:183 187.
Piomelli, S. 1981. Chemical toxicity of red cells. Environ. Health Perspect. 39:65 70.
Piomelli, S., and J. Graziano. 1980. Laboratory diagnosis of lead poisoning. Pediatr. Clin. North Am. 27:843 853.
Piomelli, S., B. Davidow, V. Guinee, P. Young, and G. Gay. 1973. The FEP [free erythrocyte porphyrin] test: A screening micromethod for lead poisoning. Pediatrics 51:254 259.
Piomelli, S., C. Seaman, D. Zullow, A. Curran, and B. Davidow.

1982. Threshold for lead damage to heme synthesis in urban children. Proc. Natl. Acad. Sci. USA 79:3335 3339.
Piomelli, S., J.F. Rosen, J.J. Chisolm, Jr., and J.W. Graef. 1984. Management of childhood lead poisoning. J. Pediatr. 105:523 532.
Piomelli, S., C. Seaman, and S. Kapoor. 1987. Lead-induced abnormalities of porphyrin metabolism. Ann. N.Y. Acad. Sci. 514:278 288.
Pirkle, J.L., J. Schwartz, J.R. Landis, and W.R. Harlan. 1985. The relationship between blood lead levels and blood pressure and its cardiovascular risk implications. Am. J. Epidemiol. 121:246 258.
Plantz, M.R., J.S. Fritz, F.G. Smith, and R.S. Houk. 1989. Separation of trace metal complexes for analysis of samples of high salt content by inductively coupled plasma mass spectrometry. Anal. Chem. 61:149 153.
Pocock, S.J., A.G. Shaper, M. Walker, C.J. Wale, B. Clayton, T. Delves, R.F. Lacey, R.F. Packham, and P. Powell. 1983. Effects of tap water lead, water hardness, alcohol, and cigarettes on blood lead concentrations. J. Epidemiol. Commun. Health 37:1 7.
Pocock, S.J., A.G. Shaper, D. Ashby, and T. Delves. 1985. Blood lead and blood pressure in middle-aged men. Pp. 303 305 in International Conference on Heavy Metals in the Environment, Vol. 1, J.D. Lekkas, ed. Edinburgh: CEP Consultants.
Pocock, S., D. Ashby, and M.A. Smith. 1987. Lead exposure and children's intellectual performance. Int. J. Epidemiol. 16:57 67.
Pocock, S.J., A.G. Shaper, D. Ashby, H.T. Delves, and B.E. Clayton. 1988. The relationship between blood lead, blood pressure, stroke, and heart attacks in middle-aged British men. Environ. Health Perspect. 78:23 30.
Pollitt, E., C. Saco-Pollitt, R.L. Leibel, and F.E. Viteri. 1986. Iron deficiency and behavioral development in infants and preschool children. Am. J. Clin. Nutr. 43:555 565.
Pollitt, E., P. Hathirat, N.J. Kotchabhakdi, L. Missell, and A. Valyasevi. 1989. Iron deficiency and educational achievement in Thailand. Am. J. Clin. Nutr. 50(3 Suppl.):687 697.
Pontifex, A.H., and A.K. Garg. 1985. Lead poisoning from an Asian Indian folk remedy. Can. Med. Assoc. J. 133:1227 1228.
Pope, A. 1986. Exposure of Children to Lead-Based Paints. EPA Contract No. 68-02-4329. Strategies and Air Standards Division.

Research Triangle Park, N.C.: U.S. Environmental Protection Agency.

Posnett, S.J., M.M. Oosthuizen, A.C. Cantrell, and J.A. Myburgh. 1988. Properties of membrane bound ferrochelatase purified from baboon liver mitochondria. Int. J. Biochem. 20:845 855.

Potluri, V.R., K.H. Astrin, J.G. Wetmur, D.F. Bishop, and R.J. Desnick. 1987. Human delta-aminolevulinate dehydratase: Chromosomal localization to 9q34 by in situ hybridization. Hum. Genet. 76:236 239.

Pounds, J.G. 1984. Effect of lead intoxication on calcium homeostasis and calcium-mediated cell function: A review. NeuroToxicology 5:295 332.

Pounds, J.G., and J.F. Rosen. 1988. Cellular Ca^{2+} homeostasis and Ca^{2+}-mediated cell processes as critical targets for toxicant action: Conceptual and methodological pitfalls. Toxicol. Appl. Pharmacol. 94:331 341.

Pounds, J.G., R.J. Marlar, and J.R. Allen. 1978. Metabolism of lead-210 in juvenile and adult rhesus monkeys (*Macaca mulatta*). Bull. Environ. Contam. Toxicol. 19:684 691.

Pounds, J.G.., G.J. Long, and J.F. Rosen. 1991. Cellular and molecular toxicity of lead in bone. Environ. Health Perspect. 91:17 32.

Preiss, J., and P. Handler. 1958. Biosynthesis of diphosphopyridine nucleotide. II. Enzymatic aspects. J. Biol. Chem. 233:493 500.

Price, J., H. Baddeley, J.A. Kenardy, B.J. Thomas, and B.W. Thomas. 1984. In vivo X-ray fluorescence estimation of bone lead concentrations in Queensland adults. Br. J. Radiol. 57:29 33.

Puchelt, H., and T. Noeltner. 1988. The stability of highly diluted multielement calibration solutions Experiences with inductively coupled plasma-mass spectrometry. Fresenius Z. Anal. Chem. 331:216 219.

Pueschel, S.M., L. Kopito, and H. Schwachman. 1972. Children with an increased lead burden: A screening and follow-up study. J. Am. Med. Assoc. 222:462 466.

Purchase, N.G., and J.E. Fergusson. 1986. Lead in teeth: The influence of the tooth type and the sample within a tooth on lead levels. Sci. Total Environ. 52:239 250.

Raab, G.M., G.O.B. Thomson, L. Boyd, M. Fulton, and D.P.H. Laxen. 1990. Blood lead levels, reaction time, inspection time and

ability in Edinburgh, Scotland, UK, children. Br. J. Dev. Psychol. 8:101 118.
Rabinowitz, M.B. 1987. Stable isotope mass spectrometry in childhood lead poisoning. Biol. Trace Elem. Res. 12:223 229.
Rabinowitz, M.B., and D.C. Bellinger. 1988. Soil lead-blood lead relationship among Boston children. Bull. Environ. Contam. Toxicol. 41:791 797.
Rabinowitz, M.B., and H.L. Needleman. 1982. Temporal trends in the lead concentrations of umbilical cord blood. Science 216:1429 1431.
Rabinowitz, M.B., and G.W. Wetherill. 1972. Identifying sources of lead contamination by stable isotope techniques. Environ. Sci. Technol. 6:705 709.
Rabinowitz, M.B., G.W. Wetherill, and J.D. Kopple. 1974. Studies of human lead metabolism by use of stable isotope tracers. Environ. Health Perspect. 7:145 153.
Rabinowitz, M.B., G.W. Wetherill, and J.D. Kopple. 1976. Kinetic analysis of lead metabolism in healthy humans. J. Clin. Invest. 58:260 270.
Rabinowitz, M.B., G.W. Wetherill, and J.D. Kopple. 1977. Magnitude of lead intake from respiration by normal man. J. Lab. Clin. Med. 90:238 248.
Rabinowitz, M.B., J.D. Kopple, and G.W. Wetherill. 1980. Effect of food intake and fasting on gastrointestinal lead absorption in humans. Am. J. Clin. Nutr. 33:1784 1788.
Rabinowitz, M., A. Leviton, and H. Needleman. 1984. Variability of blood lead concentrations during infancy. Arch. Environ. Health 39:74 77.
Rabinowitz, M., A. Leviton, and D. Bellinger. 1985a. Home refinishing: Lead paint and infant blood lead levels. Am. J. Public Health 75:403 404.
Rabinowitz, M., A. Leviton, and H. Needleman. 1985b. Lead in milk and infant blood: A dose-response model. Arch. Environ. Health 40:283 286.
Rabinowitz, M., D. Bellinger, A. Leviton, H. Needleman, and S. Schoenbaum. 1987. Pregnancy hypertension, blood pressure during labor, and blood lead levels. Hypertension 10:447 451.
Rabinowitz, M., J.D. Wang, and W.-T. Soong. 1991. Dentine lead and child intelligence in Taiwan. Arch. Environ. Health 46:351 360.

Ragaini, R.C., H.R. Ralston, and N. Roberts. 1977. Environmental trace metal contamination in Kellogg, Idaho, near a lead smelting complex. Environ. Sci. Technol. 11:773 781.
Ragan, H.A. 1977. Effect of iron deficiency on the absorption and distribution of lead and cadmium in rats. J. Lab. Clin. Med. 90: 700 706.
Raghavan, S.R.V., B.D. Culver, and H.C. Gonick. 1980. Erythrocyte lead-binding protein after occupational exposure. Environ. Res. 22:264 270.
Raghavan, S.R.V., B.D. Culver, and H.C. Gonick. 1981. Erythrocyte lead-binding protein after occupational exposure. II. Influence on lead inhibition of membrane Na^+,K^+-adenosinetriphosphatase. J. Toxicol. Environ. Health 7:561 568.
Rasmussen, H. 1986a. The calcium messenger system (Part 1). N. Engl. J. Med. 314:1094 1101.
Rasmussen, H. 1986b. The calcium messenger system (Part 2). N. Engl. J. Med. 314:1164 1170.
Regan, C.M. 1989. Lead-impaired neurodevelopment. Mechanisms and threshold values in the rodent. Neurotoxicol. Teratol. 11:533 537.
Regan, C.M. In press. Neural cell adhesion molecules, neuronal development and lead toxicity. NeuroToxicology 14(2).
Reaves, G.A., and M.L. Berrow. 1984. Total lead concentrations in Scottish soils. Geoderma 32:1 8.
Reichel, H., H.P. Koeffler, and A.W. Norman. 1989. The role of the vitamin D endocrine system in health and disease. N. Engl. J. Med. 320:980 991.
Reiter, E.R., T. Henmi, and P.C. Katen. 1977. Modeling atmospheric transport. Pp. 73 92 in Lead in the Environment, National Science Foundation Report NSF/RA-770214, W.R. Boggess, ed. Washington, D.C.: National Science Foundation.
Revis, N.W., A.R. Zinsmeister, and R. Bull. 1981. Atherosclerosis and hypertension induction by lead and cadmium ions: An effect prevented by calcium ion. Proc. Natl. Acad. Sci. USA 78:6494 6498.
Rey-Alvarez, S., and T. Menke-Hargrave. 1987. Deleading dilemma: A pitfall in the management of childhood lead poisoning. Pediatrics 79:214 217.

Richards, W.N., and M.R. Moore. 1984. Lead hazard controlled in Scottish water systems. J. Am. Water Works Assoc. 76(8):60 67.

Richardt, G., G. Federolf, and E. Habermann. 1986. Affinity of heavy metal ions to intracellular Ca^{2+}-binding proteins. Biochem. Pharmacol. 35:1331 1335.

Ritz, E., J. Mann, and M. Stoeppler. 1988. Lead and the kidney. Adv. Nephrol. Necker Hosp. 17:241 274.

Robbins, J.A. 1978. Geochemical and geophysical applications of radioactive lead. Pp. 285 393 in The Biogeochemistry of Lead in the Environment. Part A. Ecological Cycles. Topics in Environmental Health, Vol. 1A, J.O. Nriagu, ed. Amsterdam: Elsevier/North-Holland.

Roberts, T.M. 1975. A review of some biological effects of lead emissions from primary and secondary smelters. Pp. 503 532 in International Conference on Heavy Metals in the Environment, Vol. 2, Part 2, T.C. Hutchinson, ed. Toronto: Institute for Environmental Studies, University of Toronto.

Robins, J.M., M.R. Cullen, B.B. Connors, and R.D. Kayne. 1983. Depressed thyroid indexes associated with occupational exposure to inorganic lead. Arch. Intern. Med. 143:220 224.

Robinson, G.S., S. Baumann, D. Kleinbaum, C. Barton, S.R. Schroeder, P. Mushak, and D.A. Otto. 1985. Effects of low to moderate lead exposure on brain stem auditory evoked potentials in children. Pp. 177 182 in World Health Organization Environmental Health Document 3. Copenhagen: World Health Organization Regional Office for Europe.

Roda, S.M., R.D. Greenland, R.L. Bornschein, and P.B. Hammond. 1988. Anodic stripping voltammetry procedure modifed for improved accuracy of blood lead analysis. Clin. Chem. 34:563 567.

Rodamilans, M., M.J. Martinez Osaba, J. To-Figueras, F. Rivera Fillat, J.M. Marques, P. Perez, and J. Corbella. 1988. Lead toxicity on endocrine testicular function in an occupationally exposed population. Hum. Toxicol. 7:125 128.

Roels, H.A., R.R. Lauwerys, J.P. Buchet, and M.T. Vrelust. 1975. Response of free erythrocyte porphyrin and urinary δ-aminolevulinic acid in men and women moderately exposed to lead. Int. Arch. Arbeitsmed. 34:97 108.

Roels, H.A., J.P. Buchet, R. Lauwerys, G. Hubermont, P. Bruaux, F.

Claeys-Thoreau, A. Lafontaine, and J. van Overschelde. 1976. Impact of air pollution by lead on the heme biosynthetic pathway in school-age children. Arch. Environ. Health 31:310 316.

Roels, H., G. Hubermont, J.P. Buchet, and R. Lauwerys. 1978. Placental transfer of lead, mercury, cadmium, and carbon monoxide in women. Environ. Res. 16:236 247.

Rogan, W.J., J.R. Reigart, and B.C. Gladen. 1986. Association of amino levulinate dehydratase levels and ferrochelatase inhibition in childhood lead exposure. J. Pediatr. 109:60 64.

Rom, W.N. 1976. Effects of lead on the female and reproduction: A review. Mt. Sinai J. Med. 43:542 552.

Rosen J.F., and R.W. Chesney. 1983. Circulating calcitriol concentrations in health and disease. J. Pediatr. 103:1 17.

Rosen, J.F., and M.E. Markowitz. In press. Trends in the management of childhood lead poisoning. NeuroToxicology.

Rosen, J.F., and J.G. Pounds. 1989. Quantitative interactions between Pb^{2+} and Ca^{2+} homeostasis in cultured osteoclastic bone cells. Toxicol. Appl. Pharmacol. 98:530 543.

Rosen, J.F., R.W. Chesney, A.J. Hamstra, H.F. DeLuca, and K.R. Mahaffey. 1980. Reduction in 1,25-dihydroxyvitamin D in children with increased lead absorption. N. Engl. J. Med. 302:1128 31.

Rosen, J.F., M.E. Markowitz, P.E. Bijur, S.T. Jenks, L. Wielopolski, J.A. Kalef-Ezra, and D.N. Slatkin. 1989. L-line x-ray fluorescence of cortical bone lead compared with the $CaNa_2EDTA$ test in lead-toxic children: Public health implications. Proc. Natl. Acad. Sci. USA 86:685 689; Correction 86:7595.

Rosen, J.F., M.E. Markowitz, P.E. Bijur, S.T. Jenks, L. Wielopolski, J.A. Kalef-Ezra, and D.N. Slatkin. 1991. Sequential measurements of bone lead content by L X-ray fluorescence in $CaNa_2EDTA$-treated lead-toxic children. Environ. Health Perspect. 93:271 277.

Rosen, J.F., A.F. Crocetti, K. Balbi, J. Balbi, C. Bailey, I. Clemente, N. Redkey, and S. Grainger. In press. Bone lead content assessed by L-line x-ray fluorescence in lead-exposed and non-lead-exposed suburban populations in the United States. Proc. Natl. Acad. Sci. USA.

Rothenberg, S.J., L. Schnaas, C.J.N. Mendes, and H. Hidalgo. 1989. Effects of lead on neurobehavioural development in the first thirty days of life. Pp. 387 395 in Lead Exposure and Child Development:

An International Assessment, M.A. Smith, L.D. Grant, and A.I. Sors, eds. Dordrecht, The Netherlands: Kluwer Academic.

Roy, B.R. 1977. Effects of particle sizes and solubilities of lead sulfide dust on mill workers. Am. Ind. Hyg. Assoc. J. 38:327 332.

Rummo, J.H., D.K. Routh, N.J. Rummo, and J.F. Brown. 1979. Behavioral and neurological effects of symptomatic and asymptomatic lead exposure in children. Arch. Environ. Health 34:120 124.

Rush, D., N.L. Sloan, J. Leighton, J.M. Alvir, D.G. Horvitz, W.B. Seaver, G.C. Garbowsk, S.S. Johnson, R.A. Kulka, M. Holt, J.W. Devore, J.T. Lynch, M.B. Woodside, and D.S. Shanklin. 1988a. The national WIC evaluation: Evaluation of the Special Supplemental Food Program for Women, Infants, and Children. V. Longitudinal study of pregnant women. Am. J. Clin. Nutr. 48(Suppl.):439 483.

Rush, D., J. Leighton, N.L. Sloan, J.M. Alvir, D.G. Horvitz, W.B. Seaver, G.C. Garbowsk, S.S. Johnson, R.A. Kulka, J.W. Devore, M. Holt, J.T. Lynch, T.G. Virag, M.B. Woodside, and D.S. Shanklin. 1988b. The national WIC evaluation: Evaluation of the Special Supplemental Food Program for Women, Infants, and Children. VI. Study of infants and children. Am. J. Clin. Nutr. 48(Suppl.):484 511.

Russ, G.P., III. 1989. Isotope ratio measurements using ICP-Ms. Pp. 90 114 in Applications of Inductively Coupled Plasma Mass Spectrometry, A.R. Date and A.L. Gray, eds. New York: Chapman and Hall.

Russell, R.D., and R.M. Farquhar. 1960. Lead Isotopes in Geology. New York: Interscience. 243 pp.

Russell, C.S., R. Taylor, and C.E. Law. 1968. Smoking in pregnancy, maternal blood pressure, pregnancy outcome, baby weight and growth and other related factors. A prospective study. Br. J. Prev. Soc. Med. 22:119 126.

Ryu, J.E., E.E. Ziegler, S.E. Nelson, and S.J. Fomon. 1983. Dietary intake of lead and blood lead concentration in early infancy. Am. J. Dis. Child. 137:886 891.

Ryu, J.E., E.E. Ziegler, S.E. Nelson, and S.J. Fomon. 1985. Dietary and environmental exposure to lead and blood during early infancy. Pp. 187 209 in Dietary and Environmental Lead: Human Health Effects, K.R. Mahaffey, ed. Amsterdam: Elsevier.

Saenger, P., M.E. Markowitz, and J.F. Rosen. 1984. Depressed excretion of 6β-hydroxycortisol in lead-toxic children. J. Clin. Endocrinol. Metab. 58:363 367.

Samuels, E.R., and J.C. Meranger. 1984. Preliminary studies on the leaching of some trace metals from kitchen faucets. Water Res. 18:75 80.

Sanderson, N.E., E. Hall, J. Clark, P. Charalambous, and D. Hall. 1987. Glow discharge mass spectrometry A powerful technique for the elemental analysis of solids. Mikrochim. Acta 1(1 6):275 290.

Sanderson, N.E., P. Charalambous, D.J. Hall, and R. Brown. 1988. Quantitative aspects of glow discharge mass spectrometry. J. Res. Natl. Bur. Stand. (U.S.) 93:426 428.

Sandstead, H.H., D.N. Orth, K. Abe, and J. Steil. 1970. Lead intoxication: Effect on pituitary and adrenal function in man. Clin. Res. 18:76.

Sassa, S., J.L. Granick, S. Granick , A. Kappas, and R.D. Levere. 1973. Studies in lead poisoning: I. Microanalysis of erythrocyte protoporphyrin levels by spectrofluorometry in the detection of chronic lead intoxication in the subclinical range. Biochem. Med. 8:135 148.

Satzger, R.D., F.L. Fricke, P.G. Brown, and J.A. Caruso. 1987. Detection of halogens as positive ions using a helium microwave induced plasma as an ion source for mass spectrometry. Spectrochim. Acta 42B:705 712.

Savitz, D.A., E.A. Whelan, A.S. Rowland, and R.C. Kleckner. 1990. Maternal employment and reproductive risk factors. Am. J. Epidemiol. 132:933 945.

Sayre, J.W., E. Charney, J. Vostal, and I.B. Pless. 1974. House and hand dust as a potential source of childhood lead exposure. Am. J. Dis. Child. 127:167 170.

Schack, C.J., Jr., S.E. Pratsinis, and S.K. Friedlander. 1985. A general correlation for deposition of suspended particles from turbulent gases to completely rough surfaces. Atmos. Environ. 19:953 960.

Schaller, K.H., J. Gonzales, J. Thuerauf, and R. Schiele. 1980. Early detection of kidney damage in workers exposed to lead, mercury, and cadmium [in German]. Zentralbl. Bakteriol. Mikrobiol. Hyg. Abt. 1: Orig. B 171:320 335.

Schanne, F.A.X., T.L. Dowd, R.K. Gupta, and J.F. Rosen. 1989. Lead increases free Ca^{2+} concentration in cultured osteoblastic bone cells: Simultaneous detection of intracellular free Pb^{2+} by ^{19}F NMR. Proc. Natl. Acad. Sci. USA 86:5133 5135.

Schanne, F.A.X., T.L. Dowd, R.K. Gupta, and J.F. Rosen. 1990a. Development of ^{19}F NMR for measurement of $[Ca^{2+}]_i$ and $[Pb^{2+}]_i$ in cultured osteoblastic bone cells. Environ. Health Perspect. 84:99 106.

Schanne, F.A.X., T.L. Dowd, R.K. Gupta, and J.F. Rosen. 1990b. Effect of lead on parathyroid hormone-induced responses in rat osteoblastic osteosarcoma cells (ROS 17/2.8) using ^{19}F NMR. Biochim. Biophys. Acta 1054:250 255.

Schaule, B.K., and C.C. Patterson. 1981. Lead concentrations in the North East Pacific: Evidence for global anthropogenic perturbations. Earth Planet. Sci. Lett. 54:97 116.

Schaule, B.K., and C.C. Patterson. 1983. Perturbations of the natural lead depth profile in the Sargasso Sea by industrial lead. Pp. 487 503 in Trace Metals in Sea Water, NATO Conference Series IV, Vol. 9, C.S. Wong, E. Boyle, K.W. Bruland, J.S. Burton, and E.D. Goldberg, eds. New York: Plenum.

Schidlovsky, G., K.W. Jones, D.E. Burger, F.L. Milder, and H. Hu. 1990. Distribution of lead in human bone: II. Proton microprobe measurements. Basic Life Sci. 55:275-280.

Schmidt, P.F. 1984. Localization of trace elements with the laser microprobe mass analyzer. Trace Elem. Med. 1:13 20.

Schmidt, P.F., and K. Ilsemann. 1984. Quantitation of laser-microprobe-mass-analysis results by the use of organic mass peaks for internal standards. Scanning Electron Microsc. 1:77 85.

Schock, M.R., and C.H. Neff. 1988. Trace metal contamination from brass fittings. Am. Water Works Assoc. J. 80:47 56.

Schock, M.R., R. Levin, and D.D. Cox. 1989. The significance of sources of temporal variability of lead in corrosion evaluation and monitoring program design. Pp. 729 745 in Proceedings of the Water Quality Technology Conference: Advances in Water Analysis and Treatment, St. Louis, Nov. 13 17, 1988. Denver, Colo.: American Water Works Association.

Schroeder, S.R., and B. Hawk. 1987. Psycho-social factors, lead exposure, and IQ. Pp. 97 137 in Toxic Substances and Mental

Retardation: Neurobehavioral Toxicology and Teratology. Monographs of the American Association on Mental Deficiency, No. 8, S.R. Schroeder, ed. Washington, D.C.: American Association on Mental Deficiency.

Schroeder, S.R., B. Hawk, D.A. Otto, P. Mushak, and R.E. Hicks. 1985. Separating the effects of lead and social factors on IQ. Environ. Res. 38:144 154.

Schuck, E.A., and J.K. Locke. 1970. Relationship of automotive lead particulates to certain consumer crops. Environ. Sci. Technol. 4:324 330.

Schütz, A., S. Skerfving, J. Ranstam, and J.-O. Christoffersson. 1987a. Kinetics of lead in blood after the end of occupational exposure. Scand. J. Work Environ. Health 13:221 231.

Schütz, A., S. Skerfving, J.-O. Christoffersson, and I. Tell. 1987b. Chelatable lead versus lead in human trabecular and compact bone. Sci. Total Environ. 61:201 209.

Schwar, M.J., and D.J. Alexander. 1988. Redecoration of external leaded paint work and lead-in-dust concentrations in school playgrounds. Sci. Total Environ. 68:45 59.

Schwartz, J. 1988. The relationship between blood lead and blood pressure in the NHANES II survey. Environ. Health Perspect. 78:15 22.

Schwartz, J. 1991. Lead, blood pressure, and cardiovascular disease in men and women. Environ. Health Perspect. 91:71 75.

Schwartz, J. 1992a. Low Level Lead Exposure and Children's IQ: A Meta-Analysis and Search for a Threshold. Paper presented at the Annual Ramazzini Days, November 1992, Carpi, Italy.

Schwartz, J. 1992b. Lead, blood pressure, and cardiovascular disease. Pp. 223 231 in Human Lead Exposure, H.L. Needleman, ed. Boca Raton, Fla.: CRC Press.

Schwartz, J. In press. Beyond LOELs, p values, and vote counting: Methods for looking at the shapes and strengths of associations. NeuroToxicology.

Schwartz, J. In press(a). Low-level lead exposure and children's IQ: A meta-analysis and search for a threshold. Environ. Res.

Schwartz, J., and R. Levin. 1991. The risk of lead toxicity in homes with lead paint hazard. Environ. Res. 54:1 7.

Schwartz, J., and D. Otto. 1987. Blood lead, hearing thresholds, and neurobehavioral development in children and youth. Arch. Environ. Health 42:153 160.

Schwartz, J., and D. Otto. 1991. Lead and minor hearing impairment. Arch. Environ. Health. 46:300 305.

Schwartz, J., and H. Pitcher. 1989. The relationship between gasoline lead and blood lead in the United States. J. Official Stat. 5:421 431.

Schwartz, J., H. Pitcher, R. Levin, B. Ostro., and A.L. Nichols. 1985. Costs and Benefits of Reducing Lead in Gasoline: Final Regulatory Impact Analysis. EPA-230/05-85-006. Washington, D.C.: U.S. Environmental Protection Agency.

Schwartz, J., C. Angle, and H. Pitcher. 1986. Relationship between childhood blood-lead levels and stature. Pediatrics 77:281 288.

Schwartz, J., P.J. Landrigan, R.G. Feldman, E.K. Silbergeld, E.L. Baker, and I.H. von Lindern. 1988. Threshold effect in lead-induced peripheral neuropathy. J. Pediatr. 112:12 17.

Schwartz, J., P.J. Landrigan, E.L. Baker, Jr., W.A. Orenstein, and I.H. von Lindern. 1990. Lead-induced anemia: Dose-response relationships and evidence for a threshold. Am. J. Public Health 80:165 168.

Secchi, G.C., A. Rezzonico, and L. Alessio. 1968. Changes in Na^+-K^+-ATPase activity of erythrocyte membranes in different phases of lead poisoning. Med. Lav. 22:191 196.

Secchi, G.C., L. Erba, and G. Cambiaghi. 1974. Delta-aminolevulinic acid dehydratase activity of erythrocytes and liver tissue in man: Relationship to lead exposure. Arch. Environ. Health 28:130 132.

Selevan, S.G., P.J. Landrigan, F.B. Stern, and J.H. Jones. 1985. Mortality of lead smelter workers. Am. J. Epidemiol. 122:673 683.

Selevan, S.G., P.J. Landrigan, F.B. Stern, and J.H. Jones. 1988. Lead and hypertension in a mortality study of lead smelter workers. Environ. Health Perspect. 78:65 66.

Settle, D.M., and C.C. Patterson. 1980. Lead in albacore: Guide to lead pollution in Americans. Science 207:1167 1176.

Shaheen, S.J. 1984. Neuromaturation and behavioral development: The case of childhood lead poisoning. Dev. Psychol. 20:542 550.

Shapiro, I.M., and J. Marecek. 1984. Dentine lead concentration as a predictor of neuropsychological functioning in inner-city children. Biol. Trace Elem. Res. 6:69 78.

Shapiro, I.M., A. Burke, G. Mitchell, and P. Bloch. 1978. X-ray fluoresence analysis of lead in teeth of urban children in situ: Correlation between the tooth lead level and the concentration of blood lead and free erythroporphyrins. Environ. Res. 17:46 52.

Sharp, D.S., J. Osterloh, C.E. Becker, B. Bernard, A.H. Smith, J.M. Fisher, S.L. Syme, B.L. Holman, and T. Johnston. 1988. Blood pressure and blood lead concentration in bus drivers. Environ. Health Perspect. 78:131 137.

Shelton, K.R., and P.M. Egle. 1982. The proteins of lead-induced intranuclear inclusion bodies. J. Biol. Chem. 257:11802 11807.

Shelton, K.R., J.M. Todd, and P.M. Egle. 1986. The induction of stress-related proteins by lead. J. Biol. Chem. 261:1935 1940.

Shen, G.T., and E.A. Boyle. 1988. Determination of lead, cadmium, and other trace metals in annually-banded corals. Chem. Geol. 67:47 62.

Sherlock, J., G. Smart, G.I. Forbes, M.R. Moore, W.J. Patterson, W.N. Richards, and T.S. Wilson. 1982. Assessment of lead intakes and dose-response for a population in Ayr exposed to a plumbosolvent water supply. Hum. Toxicol. 1:115 122.

Sherlock, J.C., D. Ashby, H.T. Delves, G.I. Forbes, M.R. Moore, W.J. Patterson, S.J. Pocock, M.J. Quinn, W.N. Richards, and T.S. Wilson. 1984. Reduction in exposure to lead from drinking water and its effect on blood lead concentration. Hum. Toxicol. 3:383 392.

Shirahata, H., R.W. Elias, C.C. Patterson, and M. Koide. 1980. Chronological variations in concentrations and isotopic compositions of anthropogenic atmospheric lead in sediments of a remote subalpine pond. Geochim. Cosmochim. Acta 44:149 162.

Shirai, T., M. Ohshima, A. Masuda, S. Tamano, and N. Ito. 1984. Promotion of 2-(ethylnitrosamino)ethanol-induced renal carcinogenesis in rats by nephrotoxic compounds: Positive responses with folic acid, basic lead acetate, and N-(3,5-dichlorophenyl)succinimide but not with 2,3-dibromo-1-propanol phosphate. J. Natl. Cancer Inst. 72:477 482.

Shukla, R., R.L. Bornschein, K.N. Dietrich, T. Mitchell, J. Grote, O. Berger, P.B. Hammond, and P.A. Succop. 1987. Effects of fetal and early postnatal lead exposure on child's growth in stature The Cincinnati lead study. Pp. 210 212 in International Conference on Heavy Metals in the Environment, Vol. 1, S.E. Lindberg and T.C. Hutchinson, eds. Edinburgh: CEP Consultants.

Shukla, R., R.L. Bornschein, K.N. Dietrich, C.R. Buncher, O. Berger, P.B. Hammond, and P. Succop. 1989. Fetal and infant lead exposure: Effects on growth in stature. Pediatrics 84:604 612.
Shukla, R., K.N. Dietrich, R.L. Bornschein, O. Berger, and P.B. Hammond. 1991. Lead exposure and growth in the early preschool child: A follow-up report from the Cincinnati lead study. Pediatrics 88:886 892.
Shuttler, I.L., and H.T. Delves. 1986. Determination of lead in blood by atomic absorption spectrometry with electrothermal atomisation. Analyst 111:651 656.
Silbergeld, E.K. 1992. Mechanisms of lead neurotoxicity, or looking beyond the lamppost. FASEB J.6:3201 3206.
Silbergeld, E.K., R.E. Hruska, D. Bradley, J.M. Lamon, and B.C. Frykholm. 1982. Neurotoxic aspects of porphyrinopathies: Lead and succinylacetone. Environ. Res. 29:459 471.
Silbergeld, E.K., J. Schwartz, and K. Mahaffey. 1988. Lead and osteoporosis: Mobilization of lead from bone in postmenopausal women. Environ. Res. 47:79 94.
Silva, P.A, P. Hughes, S. Williams, and J.M. Faed. 1988. Blood lead, intelligence, reading attainment, and behaviour in eleven year old children in Dunedin, New Zealand. J. Child Psychol. Psychiatry 29:43 52.
Simons, T.J.B. In press. Lead-calcium interactions in cellular lead toxicity. NeuroToxicology 14(2).
Simons, T.J.B., and G. Pocock. 1987. Lead enters adrenal medullary cells through calcium channels. J. Neurochem. 48:383 389.
Skerfving, S. 1988. Biological monitoring of exposure to inorganic lead. Pp. 169 197 in Biological Monitoring of Toxic Metals, Rochester Series on Environmental Toxicity, T.W. Clarkson, L. Friberg, G.F. Nordberg, and P.R. Sager, eds. New York: Plenum.
Skoczynska, A., W. Juzwa, R. Smolik, J. Szechinski, and F.J. Behal. 1986. Response of the cardiovascular system to catecholamines in rats given small doses of lead. Toxicology. 39:275 289.
Slatkin, D.N., J.A. Kalef-Ezra, K.E. Balbi, L. Wielopolski, and J.F. Rosen. 1991. Radiation risk from L-line x ray fluorescence of tibial lead: Effective dose equivalent. Radiat. Protect. Dosimetry 37:111 116.
Slatkin, D.N., J.A. Kalef-Ezra, K.E. Balbi, L. Wielopolski, and J.F.

Rosen. 1992. L-line X ray fluorescence of tibial lead: Correction and adjustment of radiation risk to ICRP 60. Ratiat. Protect. Dosimetry 42:319 322.

Slavin, W. 1988. Atomic absorption spectrometry. Pp. 117 144 in Methods in Enzymology: Metallobiochemistry, Part A, J.F. Riordan and B.L. Vallee, eds. New York: Academic.

Smart, G.A., M. Warrington, and W.H. Evans. 1981. The contribution of lead in water to dietary lead intakes. J. Sci. Food Agric. 32:129 133.

Smith, C.M., H.F. DeLuca, Y. Tanaka, and K.R. Mahaffey. 1981. Effect of lead ingestion on functions of vitamin D and its metabolites. J. Nutr. 111:1321 1329.

Smith, D.R., S. Niemeyer, J.A. Estes, and A.R. Flegal. 1990. Stable lead isotopes evidence anthropogenic contamination in Alaskan sea otters. Environ. Sci. Technol. 24:1517 1521.

Smith, M., T. Delves, R. Lansdown, B. Clayton, and P. Graham. 1983. The effects of lead exposure on urban children: The Institute of Child Health/Southampton study. Dev. Med. Child Neurol. 25(Suppl. 47).

Somervaille, L.J., D.R. Chettle, and M.C. Scott. 1985. In vivo measurement of lead in bone using x-ray fluorescence. Phys. Med. Biol. 30:929 943.

Somervaille, L.J., D.R. Chettle, M.C. Scott, A.C. Aufderheide, J.E. Wallgren, L.E. Wittmers, Jr., and G.R. Rapp, Jr. 1986. Comparison of two in vitro methods of bone lead analysis and the implications for in vivo measurements. Phys. Med. Biol. 31:1267 1274.

Somervaille, L.J., D.R. Chettle, M.C. Scott, D.R. Tennant, M.J. McKiernan, A. Skilbeck, and W.N. Trethowan. 1988. In vivo tibia lead measurements as an index of cumulative exposure in occupationally exposed subjects. Br. J. Ind. Med. 45:174 181.

Sorrell, M., J.F. Rosen, and M. Roginsky. 1977. Interactions of lead, calcium, vitamin D and nutrition in lead-burdened children. Arch. Environ. Health 32:160 164.

Specter, M.J., V.F. Guinee, and B. Davidow. 1971. The unsuitability of random urinary delta aminolevulinic acid samples as a screening test for lead poisoning. J. Pediatr. 79:799 804.

Staessen, J. F. Sartor, H. Roels, C.J. Bulpitt, F. Claeys, G. Ducoffre, R. Fagard, R. Lauwerijs, P. Lijnen, and D. Rondia. 1991. The association between blood pressure, calcium, and other divalent cations: A population study. J. Hum. Hypertension 5:485 494.

Stark, A.D., R.F. Quah, J.W. Meigs, and E.R. DeLouise. 1982. The relationship of environmental lead to blood-lead levels in children. Environ. Res. 27:372 383.

Steele, M.J., B.D. Beck, B.L. Murphy, and H.S. Strauss. 1990. Assessing the contribution from lead in mining wastes to blood lead. Regul. Toxicol. Pharmacol. 11:158 190.

Steenhout, A., and M. Pourtois. 1981. Lead accumulation in teeth as a function of age with different exposures. Br. J. Ind. Med. 38:297 303.

Stevenson, L.G. 1949. A History of Lead Poisoning. Ph.D. Thesis. The Johns Hopkins University, Baltimore.

Stoeppler, M. 1983a. General analytical aspects of the determination of lead, cadmium and nickel in biological fluids. Pp. 133 154 in Analytical Techniques for Heavy Metals in Biological Fluids, S. Facchetti, ed. Amsterdam: Elsevier.

Stoeppler, M. 1983b. Processing biological samples for metal analyses. Pp. 31-44 in Chemical Toxicology and Clinical Chemistry of Metals, S.S. Brown and J. Savory, eds. New York: Academic.

Stoeppler, M., K. Brandt, and T.C. Rains. 1978. Contributions to automated trace analysis. Part II. Rapid method for the automated determination of lead in whole blood by electrothermal atomic-absorption spectrophotometry. Analyst 103:713 722.

Sturgeon, R., and S.S. Berman. 1987. Sampling and storage of natural water for trace metals. Crit. Rev. Anal. Chem. 18:209 244.

Sturges, W.T., and L.A. Barrie. 1987. Lead 206/207 isotope ratios in the atmosphere of North America as tracers of U.S. and Canadian emissions. Nature 329:144 146.

Succop, P.A., E.J. O'Flaherty, R.L. Bornschein, C.S. Clark, K. Krafft, P.B. Hammond, and R. Shukla. 1987. A kinetic model for estimating changes in the concentration of lead in the blood of young children. Pp. 289 291 in International Conference on Heavy Metals in the Environment, Vol. 2, S.E. Lindberg and T.C. Hutchinson, eds. Edinburgh: CEP Consultants.

Sugimoto, T., C. Ritter, I. Ried, J. Morrissey, and E. Slatopolsky. 1988. Effect of 1,25-dihydroxyvitamin D_3 on cytosolic calcium in dispersed parathyroid cells. Kidney Int. 33:850 854.

Talbot, R.W., and A.W. Andren. 1983. Relationships between Pb and Pb-210 in aerosol and precipitation at a semiremote site in northern Wisconsin. J. Geophys. Res. 88:6752 6760.
Tanner, D.C., and M.M. Lipsky. 1984. Effect of lead acetate on N-(4'-fluoro-4-biphenyl)acetamide-induced renal carcinogenesis in the rat. Carcinogenesis 5:1109 1113.
Tanquerel des Planches, L. 1839. Traite des Maladies de Plomb, ou Saturnines. Translated and edited by S.L. Dana. 1848. Lead Diseases. Lowell, Mass.: Daniel Bixby & Co.
Taylor, J.K. 1987. Pp. 7 39, 55 128, 147 171, 231 244 in Quality Assurance of Chemical Measurements. Chelsea, Mich.: Lewis Publishers.
Tepper, L.B. 1963. Renal function subsequent to childhood plumbism. Arch. Environ. Health 7:76 85.
Ter Haar, G., and R. Aronow. 1974. New information on lead in dirt and dust as related to the childhood lead problem. Environ. Health Perspect. 7:83 89.
Thomas, H.F., P.C. Elwood, E. Welsby, and A.S. St. Leger. 1979. Relationship of lead in blood in women and children to domestic water lead. Nature 282:712 713.
Thomas, H.M., and K.D. Blackfan. 1914. Recurrent meningitis due to lead in a child of five years. Am. J. Dis. Child. 8:377 380.
Thompson, J.A. 1971. Balance between intake and output of lead in normal individuals. Br. J. Ind. Med. 28:189 194.
Thompson, G.N., E.F. Robertson, and S. Fitzgerald. 1985. Lead mobilization during pregnancy [letter]. Med. J. Aust. 143:131.
Thomson, G.O., G.M. Raab, W.S. Hepburn, R. Hunter, M. Fulton, and D.P. Laxen. 1989. Blood-lead levels and children's behaviour: Results from the Edinburgh Lead Study. J. Child Psychol. Psychiatr. 30:515 528.
Tiffany-Castiglioni, E. In press. Cell culture models for lead toxicity in neuronal and glial cells. NeuroToxicology 14(2).
Todd, A.C., F.E. McNeil, J.E. Palethorpe, D.E. Peach, D.R. Chettle, M.J. Tobin, S.J. Strosko, and J.C. Rosen. 1992. In vivo x-ray fluorescence of lead in bone using K x-ray excitation with ^{109}Cd sources: Radiation dosimetry studies. Environ. Res. 57:117 132.
Trefry, J.H., S. Metz, R.P. Trocine, and T.A. Nelsen. 1985. A decline in lead transport by the Mississippi River. Science 230:439 441.

Trotter, R.T. 1985. Greta and azarcon: A survey of episodic lead-poisoning from a folk remedy. Hum. Organ. 44(1):64 72.
Turekian, K.K. 1977. The fate of metals in the oceans. Geochim. Cosmochim. Acta 41:1139 1144.
Turnlund, J.R. 1984. Trace element utilization in humans. Pp. 41 52 in Stable Isotopes in Nutrition. ACS Symposium Series 258, J.R. Turnlund and P.E. Johnson, eds. Washington, D.C.: American Chemical Society.
U.K. Central Directorate on Environmental Pollution. 1982. The Glasgow Duplicate Diet Study (1979 1980): A Joint Survey for the Department of the Environment and the Ministry of Agriculture, Fisheries and Food. Pollution Report No. 11. London: Her Majesty's Stationery Office.
Underwood, P.B., K.F. Kesler, J.M. O'Lane, and D.A. Callagan. 1967. Parental smoking empirically related to pregnancy outcome. Obstet. Gynecol. 29:1 8.
U.S. Bureau of the Census. 1985. Statistical Abstract of the United States, 1985, 105th Ed. Washington, D.C.: U.S. Department of Commerce.
U.S. Bureau of the Census. 1986. American Housing Survey, 1983. Part B: Indicators of Housing and Neighborhood Quality by Financial Characteristics. Washington, D.C.: U.S. Department of Commerce.
U.S. Bureau of the Census. 1987. Statistical Abstract of the United States, 1987, 107th Ed. Washington, D.C.: U.S. Department of Commerce.
U.S. Bureau of Mines. 1989. Mineral Commodity Summaries. Washington, D.C.: U.S. Department of Interior.
U.S. House, Committee on Energy and Commerce. 1988. Lead in Housewares. Hearing, June 27, 1988. Serial No. 100-134. Washington, D.C.: U.S. Government Printing Office.
Vahter, M., ed. 1982. Assessment of Human Exposure to Lead and Cadmium Through Biological Monitoring. Report prepared for United Nations Environment Programme and World Health Organization by National Swedish Institute of Environmental Medicine and Department of Environmental Hygiene, Karolinska Institute. Stockholm: Karolinska Institute.
Valentine, W.N., and D.E. Paglia. 1980. Erythrocyte disorders of purine and pyrimidine metabolism. Hemoglobin 4:669 681.
Valentine, J.L., R.W. Baloh, B.L. Browdy, H.C. Gonick, C.P. Brown,

G.H. Spivey, and B.D. Culver. 1982. Subclinical effects of chronic increased lead absorption A prospective study. IV. Evaluation of heme synthesis effects. J. Occup. Med. 24:120 125.

Vandecasteele, C., M. Nagels, H. Vanhoe, and R. Dams. 1988. Anal. Chim. Acta 211(1 2):91 98.

Van Loon, J.C. 1985. Selected Methods of Trace Metal Analysis: Biological and Environmental Samples. New York: Wiley-Interscience. 357 pp.

Verbueken, A.H., F.J. Bruynseels, R. Van Grieken, and F. Adams. 1988. Laser microprobe mass spectrometry. Pp. 173 256 in Inorganic Mass Spectrometry, F. Adams, R. Gijbels, and R. Van Grieken, eds. New York: John Wiley & Sons.

Veron, A., C.E. Lambert, A. Isley, P. Linet, and F. Grousset. 1987. Evidence of recent lead pollution in deep north-east Atlantic sediments. Nature 326:278 281.

Verschoor, M., A. Wibowo, R. Herber, J. van Hemmen, and R. Zielhuis. 1987. Influence of occupational low-level lead exposure on renal parameters. Am. J. Ind. Med. 12:341 351.

Victery, W., H.A. Tyroler, R. Volpe, and L.D. Grant. 1988. Summary of discussion sessions: Symposium on lead-blood pressure relationships. 78:139 155.

Victery, W., A.J. Vander, H. Markel, L. Katzman, J.M. Shulak, and C. Germain. 1982. Lead exposure begun in utero decreases renin and angiotensin II in adult rats. Proc. Soc. Exp. Biol. Med. 170: 63 67.

Victery, W., C.R. Miller, and B.A. Fowler. 1984. Lead accumulation by rat renal brush border membrane vesicles. J. Pharmacol. Exp. Ther. 231:589 596.

Vimpani, G.V., P.A. Baghurst, N.R. Wigg, E.F. Robertson, A.J. McMichael, and R.R. Roberts. 1989. The Port Pirie cohort study Cumulative lead exposure and neurodevelopmental status at age 2 years: Do HOME scores and maternal IQ reduce apparent effects of lead on Bayley mental scores? Pp. 332 344 in Lead Exposure and Child Development: An International Assessment, M.A. Smith, L.D. Grant, and A.I. Sors, eds. Dordrecht, The Netherlands: Kluwer Academic.

Waldron, H.A. 1966. The anaemia of lead poisoning: A review. Br. J. Ind. Med. 23:83 100.

Wallace, D.M., D.A. Kalman, and T.D. Bird. 1985. Hazardous lead release from glazed dinerware: A cautionary note. Sci. Total Environ. 44:289 292.

Walters, M.R., T.T. Ilenchuk, and W.C. Claycomb. 1987. 1,25-Dihydroxyvitamin D_3 stimulates $^{45}Ca^{2+}$ uptake by cultured adult rat ventricular cardiac muscle cells. J. Biol. Chem. 262:2536 2541.

Ward, N.I., R. Watson, and D. Bryce-Smith. 1987. Placental element levels in relation to fetal development for obstetrically 'normal' births: A study of 37 elements. Evidence for effects of cadmium, lead and zinc on fetal growth, and for smoking as a source of cadmium. Int. J. Biosocial Res. 9:63 81.

Wasserman, G., N. Morina, A. Musabegovic, N. Vrenezi, P. Factor-Litvak, and J. Graziano. 1991. Independent Effects of Pb Exposure and Iron Deficency on Developmental Outcome at Age Two. Paper presented at a meeting of the Society for Research in Child Development, April 1991, Seattle, Wash.

Watson, W.S., R. Hume, and M.R. Moore. 1980. Oral absorption of lead and iron. Lancet 2:236 237.

Watson, W.S., J. Morrison, M.I. Bethel, N.M. Baldwin, D.T. Lyon, H. Dobson, M.R. Moore, and R. Hume. 1986. Food iron and lead absorption in humans. Am. J. Clin. Nutr. 44:248 256.

Webb, R.C., R.J. Winquist, W. Victery, and A.J. Vander. 1981. In vivo and in vitro effects of lead on vascular reactivity in rats. Am. J. Physiol. 241:H211 H216.

Webster, R.K. 1960. Mass spectrometric isotope dilution analysis. Pp. 202 246 in Methods in Geochemistry, A.A. Smales and L.R. Wager, eds. New York: Interscience.

Wedeen, R.P. 1984. Poison in the Pot: The Legacy of Lead. Carbondale, Ill.: Southern Illinois University Press.

Wedeen, R.P. 1985. Blood lead levels, dietary calcium, and hypertension [editorial]. Ann. Intern. Med. 102:403 404.

Wedeen, R.P., J.K. Maesaka, B. Weiner, G.A. Lipat, M.M. Lyons, L.F. Vitale, and M.M. Joselow. 1975. Occupational lead nephropathy. Am. J. Med. 59:630 641.

Wedeen, R.P., D.K. Malik, and V. Batuman. 1979. Detection and treatment of occupational lead nephropathy. Arch. Intern. Med. 139:53 57.

Wedeen, R.P., P. D'Haese, F.L. van de Vyver, G.A. Verpooten, and

M.E. DeBrae. 1986. Lead nephropathy. Am. J. Kidney Dis. 8:380 383.
Weiss, S.T., A. Muñoz, A. Stein, D. Sparrow, and F.E. Speizer. 1986. The relationship of blood lead to blood pressure in a longitudinal study of working men. Am. J. Epidemiol. 123:800 808.
Whetsell, W.O., Jr., S. Sassa, and A. Kappas. 1984. Porphyrin-heme biosynthesis in organotypic cultures of mouse dorsal root ganglia: Effects of heme and lead on porphyrin synthesis and peripheral myelin. J. Clin. Invest. 74:600 607.
WHO (World Health Organization). 1977. Environmental Health Criteria 3. Lead. Geneva: World Health Organization.
WHO (World Health Organization). 1987. Lead. Pp. 242 261 in Air Quality Guidelines for Europe. European Series No. 23. Copenhagen: Regional Office for Europe, World Health Organization.
Wielopolski, L., J.F. Rosen, D.N. Slatkin, D. Vartsky, K.J. Ellis, and S.H. Cohn. 1983. Feasibility of noninvasive analysis of lead in the human tibia by soft x-ray fluorescence. Med. Phys. 10:248 251.
Wielopolski, L., J.F. Rosen, D.N. Slatkin, R. Zhang, J.A. Kalef-Ezra, J.C. Rothman, M. Maryanski, and S.T. Jenks. 1989. In vivo measurement of cortical bone lead using polarized x rays. Med. Phys. 16:521 528.
Wigg, N.R., G.V. Vimpani, A.J. McMichael, P.A. Baghurst, E.F. Robertson, and R.R. Roberts. 1988. Port Pirie cohort study: Childhood blood lead and neuropsychological development at age two years. J. Epidemiol. Commun. Health 42:213 219.
Wildt, K., R. Eliasson, and M. Berlin. 1983. Effects of occupational exposure to lead on sperm and semen. Pp. 279 300 in Reproductive and Developmental Toxicity of Metals, T.W. Clarkson, G.F. Nordberg, and P.R. Sager, eds. New York: Plenum.
Williams, H., W.H. Schulze, H.B. Rothchild, A.S. Brown, and F.R. Smith, Jr. 1933. Lead poisoning from the burning of battery casings. J. Am. Med. Assoc. 100:1485 1489.
Windebank, A.J., and P.J. Dyck. 1981. Kinetics of ^{210}Pb entry into the endoneurium. Brain Res. 225:67 73.
Winneke, G., and U. Kraemer. 1984. Neuropsychological effects of lead in children: Interactions with social background variables. Neuropsychobiology 11:195 202.
Winneke, G., K.-G. Hrdina, and A. Brockhaus. 1982. Neuropsycho-

logical studies in children with elevated tooth-lead concentrations: I. Pilot study. Int. Arch. Occup. Environ. Health 51:169 183.
Winneke, G., U. Krämer, A. Brockhaus, U. Ewers, G. Kujanek, H. Lechner, and W. Janke. 1983. Neuropsychological studies in children with elevated tooth-lead concentrations: II. Extended studies. Int. Arch. Occup. Environ. Health 51:231 252.
Winneke, G., U. Beginn, T. Ewert, C. Havestadt, U. Krämer, C. Krause, H.L. Thron, and H.M. Wagner. 1984. Study on the determination of subclinical lead effects on the nervous system of Nordenham [in German]. Schriftenr. Ver. Wasser. Boden. Lufthyg. 59: 215 229.
Winneke, G., W. Collet, U. Krämer, A. Brockhaus, T. Ewert, and C. Krause. 1987. Three- and six-year follow-up studies in lead-exposed children. Pp. 60 62 in International Conference on Heavy Metals in the Environment, Vol. 1, S.E. Lindberg and T.C. Hutchinson, eds. Edinburgh: CEP Consultants.
Winneke, G., W. Collet, and H. Lilienthal. 1988. The effects of lead in laboratory animals and environmentally-exposed children. Toxicology 49:291 298.
Winneke, G., A. Brockhaus, W. Collet, and U. Krämer. 1989. Modulation of lead-induced performance deficit in children by varying signal rate in a serial choice reaction task. Neurotoxicol. Teratol. 11:587 592.
Winneke, G., A. Brockhaus, U. Ewers, U. Krämer, and M. Neuf. 1990. Results from the European multicenter study on lead neurotoxicity in children: Implications for risk assessment. Neurotoxicol. Teratol. 12:553 559.
Wittmers, L.E., J. Wallgren, A. Alich, A.C. Aufderheide, and G. Rapp. 1988. Lead in bone. IV. Distribution of lead in the human skeleton. Arch. Environ. Health 43:381 391.
Wolff, E.W., and D.A. Peel. 1985. The record of global pollution in polar snow and ice. Nature 313:535 540.
Woods, J.S., and B.A. Fowler. 1977. Renal porphyrinuria during chronic methyl mercury exposure. J. Lab. Clin. Med. 90:266 272.
Woods, J.S., and B.A. Fowler. 1978. Altered regulation of mamalian hepatic heme biosynthesis and urinary porphyrin excretion during prolonged exposure to sodium arsenate. Toxicol. Appl. Pharmacol. 43:361 371.

Woods, J.S., D.L. Eaton, and C.B. Lukens. 1984. Studies on porphyrin metabolism in the kidney: Effects of trace metals and glutathione on renal uroporphyrinogen decarboxylase. Mol. Pharmacol. 26:336 341.

Worth, D., A. Matranga, M. Lieberman, E. DeVos, P. Karelekas, C. Ryan, and G. Craun. 1981. Lead in drinking water: The contribution of household tap water to blood lead levels. Pp. 199 225 in Environmental Lead. Proceedings of the Second International Symposium on Environmental Lead Research, D.R. Lynam, L.G. Piantanida, and J.F. Cole, eds. New York: Academic.

Wrackmeyer, B., and K. Horchler. 1989. Trimethylleadlithium in tetrahydrofuran. Synthesis of trimethyl(trimethylplumbyl)silane and of the trimethylplumbyltrihydridoborate anion [in German]. Z. Naturforsch. B: Chem. Sci. 44:1195 1198.

Yaffee, Y., C.P. Flessel, J.J. Wesolowski, A. del Rosario, G.N. Guirguis, V. Matias, T.E. Degarmo, G.C. Coleman, J.W. Gramlich, and W.R. Kelly. 1983. Identification of lead sources in California children using the stable isotope ratio technique. Arch. Environ. Health 38:237 245.

Yankel, A.J., I.H. von Lindern, and S.D. Walter. 1977. The Silver Valley lead study: The relationship between childhood blood lead levels and environmental exposure. J. Air Pollut. Control Assoc. 27:763 767.

Yip, R., T.N. Norris, and A.S. Anderson. 1981. Iron status of children with elevated blood lead concentrations. J. Pediatr. 98:922 925.

Yule, W., R. Lansdown, I.B. Millar, and M.-A. Urbanowicz. 1981. The relationship between blood lead concentrations, intelligence, and attainment in a school population: A pilot study. Dev. Med. Child Neurol. 23:567 576.

Yule, W., M.-A. Urbanowicz, R. Lansdown, and I. Millar. 1984. Teachers' ratings of children's behaviour in relation to blood lead levels. Br. J. Dev. Psychol. 2:295 305.

Zawirska, B., and K. Medras. 1972. The role of the kidneys in disorders of porphyrin metabolism during carcinogenesis induced with lead acetate. Arch. Immunol. Ther. Exp. (Warsz) 20:257 272.

Zerez, C.R., and K.R. Tanaka. 1989. Impaired erythrocyte NAD synthesis: A metabolic abnormality in thalassemia. Am. J. Hematol. 32:1 7.

Zerez, C.R., N.A. Lachant, S.J. Lee, and K.R. Tanaka. 1988. Decreased erythrocyte nicotinamide adenine dinucleotide redox potential and abnormal pyridine nucleotide content in sickle cell disease. Blood 71:512 515.

Zerez, C.R., M.D. Wong, and K.R. Tanaka. 1990. Partial purification and properties of nicotinamide adenine dinucleotide synthetase from human erythrocytes: Evidence that enzyme activity is a sensitive indicator of lead exposure. Blood 75:1576 1582.

Ziegler, E.E., B.B. Edwards, R.L. Jensen, K.R. Mahaffey, and S.J. Fomon. 1978. Absorption and retention of lead by infants. Pediatr. Res. 12:29 34.

Zielhuis, R.L., and A.A.E. Wibowo. 1976. Susceptibility of Adult Females to Lead: Effects on Reproductive Function in Females and Males [a review]. Paper presented at the Second International Workshop on Permissible Limits for Occupational Exposure to Lead, University of Amsterdam. Geneva, Switzerland: Permanent Commission, International Association of Occupational Health and Office for Occupational Health, World Health Organization.

Ziemsen, B., J. Angerer, G. Lehnert, H.-G. Benkmann, and H.W. Goedde. 1986. Polymorphism of delta-aminolevulinic acid dehydratase in lead-exposed workers. Int. Arch. Occup. Environ. Health 58:245 247.